TL 710 .D368 1989
DeLacerda, Fred.
Surviving spins

W9-ADG-439

Surviving Spins

Surviving

Spins

Fred DeLacerda

IOWA STATE UNIVERSITY PRESS / AMES

To my wife, Melissa, who made this possible
And to my flight students who make this necessary

TL
710
D368
1989

Fred DeLacerda holds degrees in engineering, mathematics, and physiology. His flight school, Delta Aviation, provides private pilot instruction as well as training in basic aerobatics and flight safety.

Unless otherwise noted, all illustrations are by Terry Polwort.

© 1989 Iowa State University Press, Ames, Iowa 50010

All rights reserved

Manufactured in the United States of America

⊗ This book is printed on acid-free paper.

No part of this book may be reproduced in any form or by any electronic or mechanical means, including information storage and retrieval systems, without written permission from the publisher, except for brief passages quoted in a review.

First edition, 1989

Library of Congress Cataloging-in-Publication Data

DeLacerda, Fred.
    Surviving spins / Fred DeLacerda.—1st ed.
        p.   cm.
    Includes bibliographical references.
    ISBN 0–8138–0142–7
    1. Airplanes—Piloting. 2. Spin (Aerodynamics) I. Title.
TL710.D368   1989
629.132′5214—dc20   89–15527

3 3001 00743 6859

CONTENTS

PREFACE

THE SPIN is a complex aspect of flight that has been a problem for the pilot since the first human became airborne. Because of its complexity, the spin has been given a mystique that is perpetuated by many stories, some of which are untrue. While some of the mysteries have been explained by undocumented trial-and-error investigations, there are well-documented research studies available that provide a more accurate understanding of the spin. Unfortunately, this theoretical knowledge is neither readily available nor easily readable for the average pilot.

There are three books available on the subject of spins: *Anatomy of a Spin* by John Lowery, *All About Stalls and Spins* by Everett Gentry, and *Stalls, Spins, and Safety* by Sammy Mason. These books discuss the stall and various spin configurations. For purposes of illustration, a variety of airplanes are discussed, ranging from the exotic (the Bede BD-5J Jet) to the antique (the OX-5 Swallow) and from the twin to the glider. Personal accounts and stories on spins are frequently used in the three books.

Why then this book? There is research data available on the normal upright spin performance of modern, light, general-aviation, single engine airplane, much of which has been collected by the National Aeronautic and Space Administration (NASA). Of equal importance is the research information on the human factor effects of a spinning airplane. The purpose of this book is to reduce this voluminous technical data into a concise form, understandable to the average general aviation pilot. This type of information is not available in the other books on this subject.

This book is not meant to be an academic textbook for theoretical discussion of spins. Also, it is not meant to be an entertaining collection of stories about spinning a variety of airplanes. Besides those airplanes tested by NASA, one airplane — the Cessna 150 — is singled out for discussion simply because of its popularity as a primary flight trainer and because of the available test data on spins.

This book is meant to be a source of factual information on the normal upright spin of modern, light, general-aviation, single-engine airplanes. The information is intended to be used by pilots, particularly students and flight instructors, in conjunction with primary flight instruction.

ACKNOWLEDGMENTS

NO HUMAN ENDEAVOR is the accomplishment of one person. It is, instead, the combined efforts of many. This book is no different, and the contributions from others span many years.

My initial interest in spins began with my first flight instructor, Bob Gilbert, who, as part of my private pilot training, introduced me to the spin in a J-3 Cub.

The interest was nurtured by Bob Robertson as he guided me through my flight training to the flight instructor certification.

Encouragement to begin the book came from the late Dr. William Kirkham, FAA Civil Aeromedical Institute, who impressed on me the need to inform general aviation pilots of human factors essential for flight safety.

Stimulation to explore every source of information came from my flight and ground instruction with William K. Kershner.

James Patton and Paul Stough of NASA assisted me in my research efforts.

Terry Polwort, a former flight student, put my ideas into illustrations and assisted me with manuscript preparation.

To each of these people I am most appreciative and forever grateful.

Surviving Spins

1
Introduction

The Problem

There is no topic of discussion in the field of aviation that generates more controversy than whether or not spin training should be required as part of pilot certification. In 1949 the Civil Aeronautics Administration (CAA) adopted CAR Amendment 20–3 thereby eliminating spin training from pilot certification requirements. It was theorized that the design of stall/spin-proof airplanes would make the spin requirement obsolete. Although a stall/spin-proof airplane has not been realized, the National Aeronautics and Space Administration (NASA) is investing effort in an attempt to accomplish such an airplane design. Meanwhile, the pros and cons of spin training continue to be argued.

Presently there is great diversity in spin training. In some cases there is no discussion of spins during any part of flight training, while in other cases there is brief ground instruction with the possibility of some flight instruction. By contrast, there are some cases where in-depth training is provided. There are those persons who firmly believe that failure to provide spin training compromises safety. There are others who are equally as resolved that training in stall awareness and stall avoidance is adequate. As in any controversy, both sides of the issue have merit.

Statistics

As with most issues involving safety, accident statistical data are used to support both positions. However, one must be reminded that Mark Twain once said there are three kinds of lies: lies, damn lies, and statistics. Keep this in mind as three sources of statistical data are examined.

One source of statistical data on spin related accidents is the Society of Automotive Engineers (SAE Paper 760480). This data is given in Table 1.1. The results indicate that most spin accidents occur in connection with aerobatics and buzzing at low altitudes although reasons for a significant number are unknown. There is essentially no difference between the number of takeoff and landing spin accidents.

Table 1.1.   Stall/Spin Accidents, 1965–1973[a]

Phase of Flight	Percentage
Takeoff	26
In Flight	
Climb/cruise/descent	8
Aerobatic/buzzing	18
Unknown	22
Landing	
Pattern	5
Final	11
Go-around	9
Unknown/other	1

[a]"Statistical Analysis of General Aviation Stall/Spin Accidents," Society of Automotive Engineers Paper 760480 (Reprinted with permission of the Society of Automotive Engineers, Inc.)

An aviation publication, *Aviation Safety,* presented some interesting statistics regarding fatal stall/spin accidents. The analysis involved approximately 3800 accidents of all types for the year 1979. The percentages of all fatal accidents were determined for 1500 pilots who were grouped according to number of flight hours: novice (less than 100 hours), fledgling (101–200 hours), journeyman (1001–1500 hours), master (5001–10,000 hours). The top three causes of fatal accidents were tabulated for each category. The percentage of fatal stall/spin accidents increased as flight time increased. Fatal stall/spins increased in rank from third for the novice to first for the journeyman and master. (Table 1.2) Interestingly, the novice category had the lowest percentage (8 percent) whereas the journeyman category had the highest (24 percent), followed closely by the master group (22 percent). It appears that experience grants no protection against a fatal stall/spin accident.

The author studied the National Transportation Safety

Table 1.2. Top Three Killers: Type of Accident as % of Fatal Accidents[a]

Novice	
VFR-into-IFR	40%
Buzzing, low flying	16%
Stall spin	8%
Fledgling	
VFR-into-IFR	50%
Stall spin	19%
Buzzing, low flying	10%
Journeyman (Ag ops excluded)	
Stall/spin	24%
VFR-into-IFR	14%
Buzzing, low flying	13%
Master (Ag ops excluded)	
Stall/spin	22%
Engine failure	18%
Improper IFR	12%

[a]"Staying Alive in an Airplane: A Game of Chance or Skill," *Aviation Safety,* vol. III, No. 7, July 1983. (Reprinted by permission from Aviation Safety, 1111 East Putnam Avenue, Riverside, Conn. 06878. All rights reserved.)

Board (NTSB) accident data for instructional flying for the period 1978 to 1983. Tabulating only those accident types given as spins the results are shown in Table 1.3. The largest number of accidents took place during an unknown phase of flight. Aerobatics accounted for the next largest group followed closely by traffic pattern operations.

It was interesting to note that in 58 percent of these accidents a certified flight instructor was on board. The following are comments related to some of the accidents.

- Center of gravity (CG) aft for utility category
- Certified flight instructor (CFI) recovered too low
- Overgross
- CFI failed to recover from intentional spin, flaps down
- Difficulty in recovery, 6 turn, prop stopped

These comments suggest a lack of awareness of critical factors essential to the execution of and recovery from a spin.

Interpretation of statistical data is relative to a person's opinion regarding spin training. It can be argued that most spin accidents start at altitudes too low for recovery; therefore, stall awareness is more critical. On the other hand, the majority of

Table 1.3. Instruction Flying Spin Accident, 1978–1983

Flight Phase	Percentage
Emergency	
Actual	3
Simulated	8
Pattern	
Take off/go around	7
Approach/landing	8
Flight maneuvers	
Practice stalls	4
Intentional spin	7
Aerobatics	19
Other	
Low flight/buzzing	9
Unknown	35

Source: author's research.

spin accidents are in the "unknown" category, hence no one can ascertain if spin training would have been of any value.

The Civil Aeronautics Board's 1949 decision to eliminate spin training and concentrate instead on stall/spin avoidance has resulted in a reduction in stall/spin-related accidents. In the period from 1945 through 1948, the time prior to elimination of required spin training, 48 percent of all fatal accidents were stall/spin-related. For the period 1967 through 1969, the number was reduced to 22 percent. According to the NTSB data for the period 1972 through 1977, stall/spin was the primary factor in 12 percent of all fatal accidents. Therefore, the effectiveness of the shift from to spin training to spin avoidance training is supported by the decline in percentage of fatal accidents due to stall/spin.

Purpose

It is apparent that endless comparative arguments, supported by statistical data, can be made. There appears to be only one inescapable conclusion. Spin training is a complex issue for which there is no single solution. Most importantly, spin training involves more than flight instruction. Any flight instruction in spin entry and recovery must be accompanied by ground instruction in all aspects of the aerodynamic spin.

Therein lies the purpose of this book. This book is intended to be factual data about spinning light, general-aviation airplanes. The contents of the book are based on the published information related to normal upright spins. The information was selected to cover all aspects of spins from aerodynamics to airplane categories. The book does not dwell on hangar stories or spin accounts by individuals.

This book is intended to be used for instructional purposes. It is not intended as a replacement for flight training but as a supplement to be used with ground instruction accompanied by flight instruction.

2
Regulations Applicable to Spins

Regulations

John Doe was a typical pilot. For several years after receiving his private license he had flown a variety of rented airplanes. Recently he had became a partner with three other pilots in ownership of a Honey Bee 123, a four-place, single-engine airplane appropriate for both business and pleasure use. With John as pilot-in-command, the four pilots went up for their first flight to explore the flight characteristics of the airplane. Finding the Honey Bee 123 easy to fly, John decided to try a spin entry and recovery. The airplane entered a spin mode from which recovery was not possible. The airplane was certified in the utility category, but there was a placard against intentional spins with passengers in the rear seats.

As a student pilot John's flight instructor had given him instruction in spin entry and recovery. The instruction had included a brief ground session on spin entry and recovery followed by approximately an hour of flight instruction in a two place trainer. Since that time John had on occasion intentionally entered and recovered from one-turn spins in a two-place trainer aircraft. Consequently, John felt comfortable with spins. Tragically, his feeling of competency was based on superficial knowledge of a complex subject.

This story is fiction but, unfortunately, the scenario is familiar. It is imperative that instruction in spin includes more than just entry and recovery. One area that needs be included is the applicable Federal Aviation Regulations, the FARs, with particular attention to airplane spin certification requirements.

Aerobatic Flight

One cannot go to a single regulation to find applicable rules for performing spins; however, there are several areas that directly affect such training. FAR 91.71 describes aerobatic flight as an intentional maneuver involving an abrupt change in the airplane's flight attitude. This includes abnormal attitudes and accelerations not needed for normal flight. As an aerobatic maneuver, spins must be done in compliance with FAR 91.71. Spins are prohibited over congested areas, over an open-air assembly of persons, within control zones or airways, below an altitude of 1500 feet above ground level (AGL), and with flight visibility of less than 3 miles.

Parachute

FAR 91.15 is more specific regarding spins. An approved parachute is required by each occupant when any intentional maneuver exceeds a bank of 60 degrees or a pitch attitude of 30 degrees. A parachute is not required for spins and other flight maneuvers *required* by regulation for any certificate or rating when given by a certificated flight instructor. Since spins are not required for private or commercial pilot certifications, parachutes must be worn during spin training.

The FAA issued an amendment (Amendment 91.6) to relieve the requirement for wearing parachutes during certain aerobatic flight. FAR 91.71(b) prohibits intentional aerobatic flight unless all occupants, other than crewmembers, are wearing parachutes. A student was not considered a crewmember, but a passenger. Hence a parachute was required for spin training. The regulation was amended to state that regardless of what certificate or rating the pilot applicant was seeking, an aerobatic maneuver required for any pilot certification may be performed without a parachute. This is interpreted to mean that even though a student pilot is not required to have flight instruction in spins his flight instructor, if he considers spin training necessary, may give the student this training without parachutes since spin instruction is required for certain other certificates. However, the only certificate requiring spin instruction is that for flight instructor.

This amendment was adopted in response to complaints from the field that a parachute, when worn within the confines of certain cockpits, may reduce the pilot's visibility and hinder manipulation of the controls. Interestingly nothing was said about alternatives if the airplane failed to recover from a spin and the pilot had no parachute. In addition, in adopting the amendment it was presumed that the flight instructor had the skill necessary to safely give aerobatic instruction required by the pilot regulations *within the operating limitations of the airplane* without the necessity of the instructor or the student wearing a parachute.

To add to the confusion, according to FAR 61.45 a flight instructor applicant need only have a logbook endorsement by a certified flight instructor that the applicant has been given spin training. With this endorsement an applicant need not demonstrate a spin on the flight test. It is therefore possible to have a flight instructor who has done nothing more than perform a one-turn spin to the left and to the right. *This* is presumed adequate preparation for safely giving spin training to a student.

Airplane Category

FAR 23.221 specifies airworthiness standards for type certification concerning spin maneuvers. A normal category airplane must be able to recover from a one-turn spin or a three-second spin, whichever takes longer, in not more than one additional turn with control input in a manner normally used for recovery. For both the flap-retracted and flap-extended conditions, the applicable airspeed limits and positive limit load (+3.8 G) must not be exceeded. There may be no excessive back pressure during spin recovery. It must be impossible to obtain uncontrollable spins with any use of the controls. There must be a visible placard prohibiting aerobatic maneuvers including spins (Fig. 2.1). A normal category airplane is permitted any maneuver

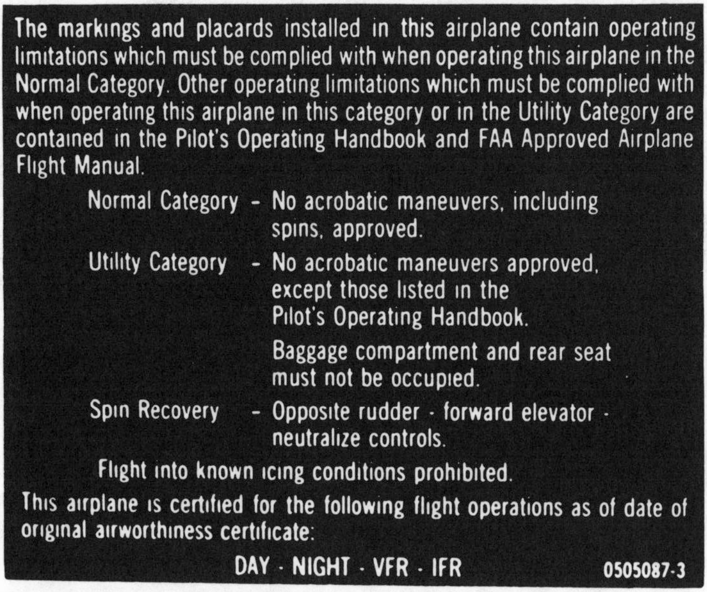

The markings and placards installed in this airplane contain operating limitations which must be complied with when operating this airplane in the Normal Category. Other operating limitations which must be complied with when operating this airplane in this category or in the Utility Category are contained in the Pilot's Operating Handbook and FAA Approved Airplane Flight Manual.

Normal Category	- No acrobatic maneuvers, including spins, approved.
Utility Category	- No acrobatic maneuvers approved, except those listed in the Pilot's Operating Handbook.
	Baggage compartment and rear seat must not be occupied.
Spin Recovery	- Opposite rudder - forward elevator - neutralize controls.

Flight into known icing conditions prohibited.

This airplane is certified for the following flight operations as of date of original airworthiness certificate:

DAY · NIGHT · VFR · IFR   0505087-3

FIG. 2.1. Placard for an airplane certified in both normal and utility categories. (Photo by author)

incident to normal flight within +3.8 and −1.52 G load. This includes lazy eights, chandelles, stalls, and steep turns that do not exceed 60 degrees angle of bank.

An aerobatic category airplane must recover from any point in a spin in no more than one and one-half additional turns after normal recovery procedures. This applies to spins of six turns or three seconds, whichever takes longer, with flaps retracted, and one turn or three seconds, whichever takes longer, with flaps extended. With flaps either retracted or extended the applicable airspeed limit and the positive limit load (+6.0 G) may not be exceeded. If a placard prohibits intentional spins

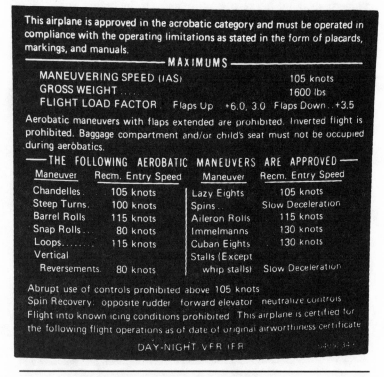

FIG. 2.2. Placard for an airplane certified
in the aerobatic category. (Photo by author)

with flaps extended, flaps may be retracted during recovery. It must be impossible to obtain uncontrollable spins with any use of the controls. There must be a visible placard listing approved maneuvers with entry speeds (Fig. 2.2). The airplane is to be operated within the +6.0 and −3.0 G-load limits.

A utility category airplane must meet the requirements for either the aerobatic or the normal category. Certification in the utility category means the airplane has been shown to recover from a six-turn spin within one and one-half turns; however, it is possible to certify an airplane in this category yet do the spin testing required in the normal category. Consequently, utility category does not necessarily mean the airplane is approved for multiturn spins. A one-turn spin test is actually an investigation of the airplane's characteristics in a prolonged stall rather than a true spin. One must be reminded that a one-turn spin is the incipient phase of a spin, and, as an entry condition, does not exhibit the characteristics of a developed spin. Therefore, a utility airplane must be spun only as approved. The utility category airplane can perform all operations in the normal category but within a load limit of +4.4 and −1.76 G. In addition to lazy eights and chandelles, steep turns of more than 60 degrees bank angle are approved.

The normal and utility category airplanes may not have been tested in the multiturn phase of a developed spin, but an airplane can be certified in both normal and utility categories. These airplanes are restricted to center of gravity envelopes for each category. This in turn determines the spin certification. The airplane must have a placard stating the restrictions for spins (Fig. 2.1). This usually restricts the center of gravity envelope to the next higher category.

FAR 23 provides for certification of spin-proof airplanes. This option is seldom used simply because compliance with the regulation would make airplane certification more difficult and thereby add to the cost of certification. Consequently, the easier certification requirement of recovery from a one-turn spin within one additional turn is used in order to reduce time and

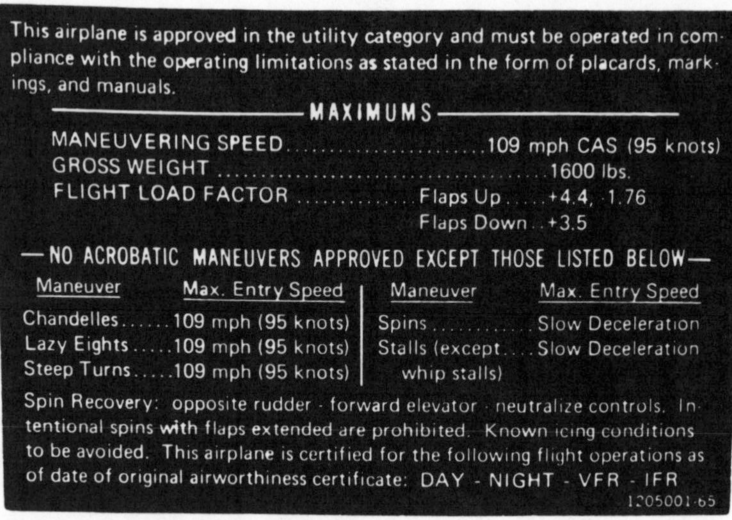

This airplane is approved in the utility category and must be operated in compliance with the operating limitations as stated in the form of placards, markings, and manuals.

——————————— MAXIMUMS ———————————

MANEUVERING SPEED............................109 mph CAS (95 knots)
GROSS WEIGHT ....1600 lbs.
FLIGHT LOAD FACTOR ....Flaps Up....+4.4, -1.76
                                       Flaps Down..+3.5

—NO ACROBATIC MANEUVERS APPROVED EXCEPT THOSE LISTED BELOW—

Maneuver	Max. Entry Speed	Maneuver	Max. Entry Speed
Chandelles......109 mph (95 knots)		Spins ....Slow Deceleration	
Lazy Eights.....109 mph (95 knots)		Stalls (except....Slow Deceleration	
Steep Turns.....109 mph (95 knots)		whip stalls)	

Spin Recovery: opposite rudder - forward elevator - neutralize controls. Intentional spins with flaps extended are prohibited. Known icing conditions to be avoided. This airplane is certified for the following flight operations as of date of original airworthiness certificate: DAY - NIGHT - VFR - IFR

1205001-65

FIG. 2.3. Placard for an airplane certified in the utility category. (Photo by author)

cost, but this produces a spinnable airplane. In essence, although spin instruction was dropped from flight training in anticipation of a spin-proof airplane, regulations on certification discourage development of such airplanes.

Older design airplanes such as the J-3 Cub and Aeronca 7-AC were built under regulations with no operational categories. These airplanes do not have category placards. Up to a weight of approximately 4000 pounds, the load limits are comparable to the utility category. Above this weight the load limits decrease as airplane weight increases and are comparable to the normal category.

Summary

While FARs may be a bore to read, it must be emphasized that intentional spins exceeding the certification requirements of the airplane make the pilot, whether student or CFI, a test pilot since the airplane has entered a flight realm for which it has not been tested. Be reminded that the certification pilot was experienced and flew a prototype airplane. The pilot was wearing a parachute and the airplane's controls were rigged within manufacturer tolerances.

The FARs pertaining to use of parachutes during spin training are confusing. Despite amendatory language to clarify the regulation, there is still considerable disagreement in the interpretation. Remember, the spin is an aerobatic maneuver for which failure to recover is fatal. Perhaps the Walt Disney character Jimminy Cricket said it best, "Let your conscience be your guide."

3
Definition of a Spin

Historical Background

Since the beginning of manned flight, stalls and spins have resulted in many crashes and fatalities. During early flight the importance of control at low speeds was underrated, hence, early airplane designs were deficient in low speed controllability. For example, Otto Lilienthal failed to provide adequate pitch control in his manned glider designs. It was Orville Wright who discovered that many nose-down crashes were due to stalls, and, he recognized that forward elevator movement was the appropriate stall recovery procedure. This was not a readily acceptable procedure for a pilot who was descending toward the earth in a nose-down attitude.

Because early flights were at low altitudes the progression of a stall into a spin was not immediately recognized. However, as manned flight moved to higher altitudes, stall progression to a spin became a significant problem. The helical path of an airplane during a spin led to the name "spiral." And, since a crash frequently ended the spin, the spin became known as the "graveyard spiral." In 1912 Wilfred Parke recognized the recovery technique for a spin, and it became known as "Parke's dive."

Airplanes used in World War I were structurally capable of spinning and recovering without being destroyed in the process. The spin was even used in aerial combat. After the war, the Curtiss "Jenny" trainer was used by the National Advisory Committee for Aeronautics (NACA) to investigate high lift conditions leading to stalls and spins. These were the first systematic flight tests in the area of airplane control at high lift conditions. The results helped explain the conditions that lead to a spin.

Relative Definition of a Spin

In the past spins were difficult to identify and name. Thus they are also difficult to define appropriately. The definition of a spin is relative to one's perspective. Consider the following:

- From the pilot perspective a spin is a rapid helical or corkscrew descent about a vertical axis. However, a pilot's reaction to his first spin could be expressed in several explicit four letter words.
- According to the Federal Aviation Agency (FAA) *Flight Training Manual* the spin is an aggravated stall that results in autorotation. The flight path is a downward corkscrew with one wing producing lift while the airplane is being forced downward by gravity as the airplane rolls and yaws.
- From the aerodynamic viewpoint the spin is flight motion at some angle of attack between stall and 90 degrees with the airplane descending toward the earth while rotating about a vertical axis. The spinning motion involves simultaneous roll, yaw, and pitch while the airplane is at a high angle of attack and sideslipping.

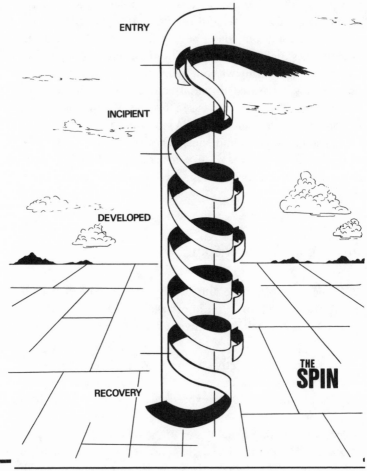

ENTRY

INCIPIENT

DEVELOPED

RECOVERY

THE
SPIN

FIG. 3.1. Four phases of a spin.

Phases of a Spin

A concise definition of a spin is complicated because it is made up of four separate phases (Fig. 3.1). The following distinctions need to be made:

1. **Entry.** The airplane is stalled in uncoordinated flight.
2. **Incipient.** This is the region following entry into the spin. While the airplane is stalled the vertical and rotational velocities are increasing.
3. **Developed.** There is self-sustaining motion of the airplane about the vertical axis. The airplane is in an equilibrium.
4. **Recovery.** Control inputs disrupt the equilibrium by stopping rotation and breaking the stall.

Summary

There is no simple definition of a spin because it is a complex subject. Superficial knowledge about spins is dangerous. To be safe a pilot must respect spins. This respect can only be acquired through a thorough understanding of all factors relative to the aerodynamic spin.

4
Aerodynamics of a Spin

Definition

Aerodynamically, a spin is an aggravated stall that results in autorotation such that the airplane descends in a helical path. This downward rotation is about an axis perpendicular to the earth's surface. The airplane is rotating about the three axes so there is combined pitch, roll, and yaw motions. The conditions of roll, yaw, and pitch are in equilibrium sustained by prospin and antispin moments. The equilibrium is such that the airplane is at a high angle of attack while in a sideslip. The air flow is in the region beyond the stall so that aerodynamic characteristics of the airplane are both nonlinear and time dependent.

Lift Coefficient

To the aerodynamicist the information above is simple, but to the average pilot it is staggering to say the least, but it is comprehensible if one takes it in small portions. Since the spin is autorotation beginning from a stall, we start with the stall. And understanding of the stall requires a working knowledge of lift coefficient. The lift coefficient is a measure of lift generated by a given airfoil. For ease in understanding the concept of lift coefficient, lift is plotted graphically relative to angle of attack (angle between wing chord and relative wind). The result is the lift curve (Fig. 4.1).

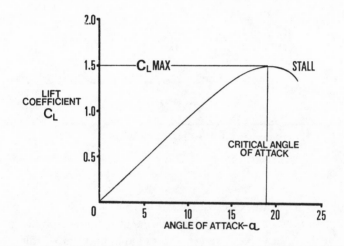

FIG. 4.1. Lift curve for an airfoil section.

The slope of the lift curves for all airfoil sections is essentially the same. For low angles of attack the coefficient of lift increases approximately 0.1 for each 1 degree increase in angle of attack. The shape of the lift curves varies for different airfoils as the maximum value for lift coefficient is approached. Compare the lift curves for the three airfoils—NACA 63–006, NACA 63–009, and NACA 63–012 (Fig. 4.2).

Note that the maximum coefficient of lift occurs at different critical angles of attack for the three airfoils. Increasing the angle of attack beyond the critical value results in a decreased

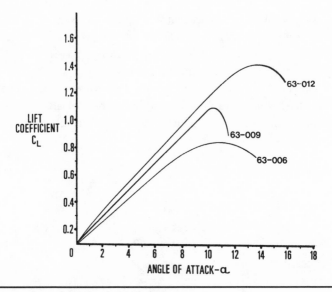

FIG. 4.2. Lift curves for three NACA airfoil sections.

coefficient of lift. Consequently, each airfoil stalls at different angles of attack. The maximum lift coefficient corresponds to a minimum flight speed, referred to as minimum controllable airspeed. This stall speed in level flight is related to the coefficient of lift as follows:

$$V = 17.2 \left[\frac{W}{CS\sigma}\right]^{\frac{1}{2}}$$

V = stall speed (knots TAS)
W = gross weight (lbs)
C = maximum coefficient of lift
σ = altitude density ratio
S = wing area (sq. ft)
17.2 = numerical constant for conversion of velocity to units of knots

For example, the stall speed of a typical two-place trainer can be determined. The maximum lift coefficient for an NACA 2412 airfoil is 1.6 at an angle of attack of 15 degrees. With a wing area of 160 square feet and a gross weight of 1600 pounds operating at 4000 feet (altitude density ratio of 0.888), the stall speed is 46 knots.

The shape of the lift curve in the region of maximum lift coefficient reflects the potential for spin entry at stall. A sharp curve top indicates a rapid decrease in lift with increasing angle of attack beyond the critical value. This decrease is due to large areas of airflow separation. Static directional stability deteriorates at high angles of attack and, when combined with the large areas of airflow separation over the wing, makes roll motion highly probable. Any roll motion at this point will be accompanied by relatively large increases in drag that may produce sufficient yaw for spin entry. The differences in shape of the lift curve in the region of maximum lift can be seen for the three NACA lift curves shown in Figure 4.2.

Spin Entry

The basic determinant for a spin requires that the airplane be placed at an angle of attack beyond the critical value. As a rule, the general aviation airplane must be stalled before a spin can take place; however, a stall in itself does not necessarily result in a spin. There must be rotation. This rotation results from asymmetry in lift and drag at stall, the asymmetry being due to airplane design or rigging, control input by the pilot, or turbulence. Spin entry from straight and level flight configuration is shown in Fig. 4.3.

At stall the airplane pitches downward. A roll displacement to the left is produced by airplane design or rigging, control input, or turbulence. With the angle of attack above stall the right upgoing wing has a decreasing angle of attack while the left downgoing wing has an increasing angle of attack. The right

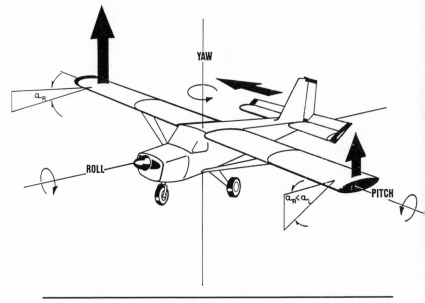

FIG. 4.3. Forces and moments acting on an airplane at spin entry.

upgoing wing becomes less stalled as the left downgoing wing becomes more stalled.

Different pitch, roll, and yaw velocities are involved as the airplane enters the spin. The left downgoing wing has a yaw velocity greater than that of the right upgoing wing. The roll velocity increases the angle of attack for the left downgoing wing and decreases the angle of attack for the right upgoing wing. The increased angle of attack for the left downgoing wing results in a decreased coefficient of lift and an increased coefficient of drag. The decreased angle of attack for the right upgoing wing gives an increased coefficient of lift and a decreased coefficient of drag. These aerodynamic features are shown in Fig. 4.4.

The greater lift of the right wing as compared to the left wing adds to the initial roll displacement. The greater drag on

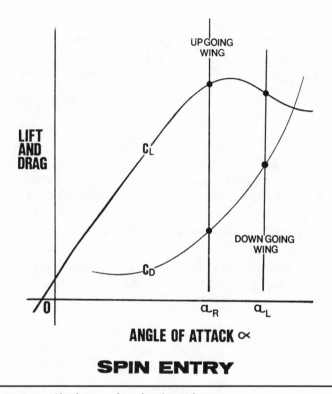

SPIN ENTRY

FIG. 4.4. Lift, drag, and angle of attack at spin entry.

the left wing as compared to the right produces yaw to the left. Therefore, due to the differential in lift the airplane rolls to the left, and, due to the differential in drag, yaws to the left. The result is rotation to the left about a vertical axis.

Autorotation

Autorotation is a self-sustaining motion whereby pitch, roll, and yaw are in equilibrium due to a balance in prospin and antispin moments. These moments are both of aerodynamic and inertial origin. To understand autorotation one must have a working understanding of moments. While pilots are knowledgeable about moments with respect to calculating weight and balance, this knowledge does not usually extend to moments involving aerodynamic and inertial forces.

MOMENT. Moment of a force about a given axis (A) measures the tendency of the force (F) to cause rotation of the airplane about the axis (Fig. 4.5). Airflow acting on the deflected rudder produces rotation of the airplane about the vertical axis acting through the center of gravity. The product of the force acting on the rudder and the perpendicular distance (D) from the rudder to the vertical axis is the aerodynamic moment. The force acting on the rudder is a function of dynamic air pressure ($1/2\varrho V^2$ where ϱ = air density and V = velocity) and surface area of the rudder. The greater the force and/or the distance, the greater the moment.

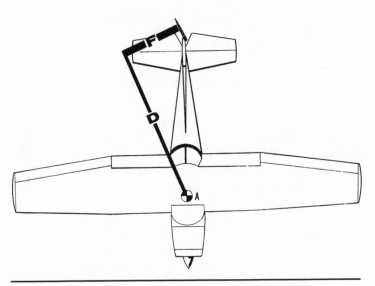

FIG. 4.5. Aerodynamic moment (F × D) produced by deflection of the rudder.

INERTIA. If a mass (M) is located a given distance (R) from an axis (B) then the product of the distance squared times the mass gives a measure of the inertia of the airplane (Fig. 4.6). Inertia is the resistance of the airplane to motion or a change in motion. Adding mass to the tail of an airplane changes the inertia. Suppose a modified engine is put in an airplane. In order to maintain the center of gravity within the envelope the battery is placed in the tail of the airplane. The moment of inertia about the vertical and lateral axes has increased. Thus rotation about either axes will be more difficult to change.

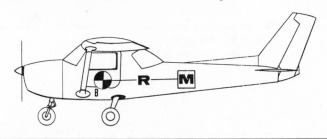

FIG. 4.6. Moment of inertia (M × R)
produced by adding a mass (M) in the tail of
an airplane.

COUPLE. The moment of a couple involves two forces having the same magnitude, parallel lines of action, and opposite directions. While the sum of the forces is zero, the moment of the forces is the product of the magnitude (F) times the perpendicular distance (D) to the axis of rotation. Thus the forces rotate but do not translate the airplane (Fig. 4.7). Couples in airplanes are formed both by aerodynamic and inertial forces. Inertial couples can be antispin or prospin, depending on the mass distribution of the airplane (Figs. 4.8 and 4.9).

PROSPIN AND ANTISPIN MOMENTS. Airflow acting on the airplane control surfaces — the ailerons, rudder, and elevator — produces aerodynamic forces. The magnitude of the aerodynamic forces is equal to the dynamic pressure of the air times the

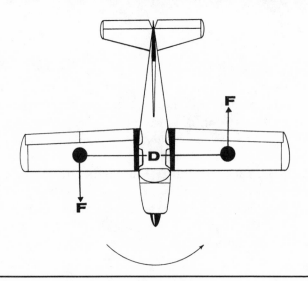

FIG. 4.7. A couple formed by location of
fuel tanks in the wings of an airplane.

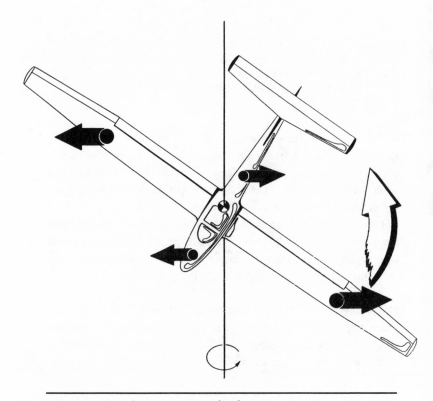

FIG. 4.8. Prospin moment (couple) due to
wings being proportionally greater than the
fuselage.

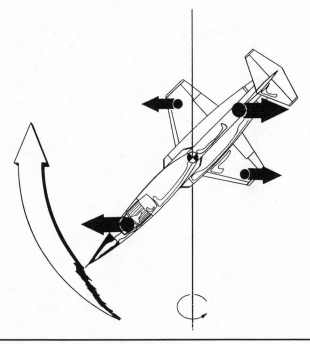

FIG. 4.9. Antispin moment (couple) due to
fuselage being proportionally greater than
the wing.

surface area of the control surfaces. The product of the aerodynamic forces times the perpendicular distance from the center
of gravity to the control surfaces creates aerodynamic moments.

In addition, the airplane is composed of mass segments
distributed throughout the airplane; therefore, the airplane has
inertia. This inertia opposes change in motion of the airplane;
hence aerodynamic moments must overcome inertia in order to
enter a spin. Once in the spin, autorotation results from a balance between the inertial and aerodynamic moments. These moments can be classified as prospin and antispin moments. Prospin moments favor rotation and antispin moments oppose
rotation.

30

EQUILIBRIUM. To fully comprehend the aerodynamics of a spin consider a pilot entering an intentional spin from level flight. The airplane is brought to a stall by up-elevator control input. The aerodynamic force on the tail produces an aerodynamic moment that pitches the nose up until reaching the critical angle of attack. Upon entering the stall the loss of lift causes a pitch-down moment. On entering the stall the rudder is deflected to the left. The aerodynamic force on the rudder produces an aerodynamic moment that yaws the airplane to the left. There is now sideslip to the right, the right wing moving upward and the left wing downward. The unequal aerodynamic forces from the unequal lift and drag create roll and yaw moments to the left (Fig. 4.10).

The inertia of the wings produces a couple that causes rotation to the left. This wing inertia couple is opposed by the fuselage inertia couple. The aerodynamic pitching moment downward is opposed by the inertial pitching moment upward. When the rotational couples are in equilibrium and the pitching moments are in equilibrium a spin rate is obtained that gives a self-sustaining motion. This is autorotation found in a steady-state, developed spin (Figure 4.11). The prospin moments and the antispin moments are in equilibrium.

TORQUE. There are several factors inherent in airplane design that produce a turning or yawing of the airplane. These factors include torque reaction of the engine, slipstream (propwash) rotation, asymmetrical propeller thrust (P-factor), and gyroscopic precession. Collectively these four factors are frequently referred to as torque. Two of these factors, gyroscopic precession and P-factor, may influence the characteristics of a spin.

Gyroscopic precession is the reaction of a rapidly spinning body such as the disc formed by a rotating propeller. When a force is applied to the rotating propeller disc, the disc reacts as if the force was exerted 90 degrees from the point of application. For a clockwise rotating propeller (as seen from the cockpit) there is a left yaw reaction when a force is applied at the twelve o'clock position.

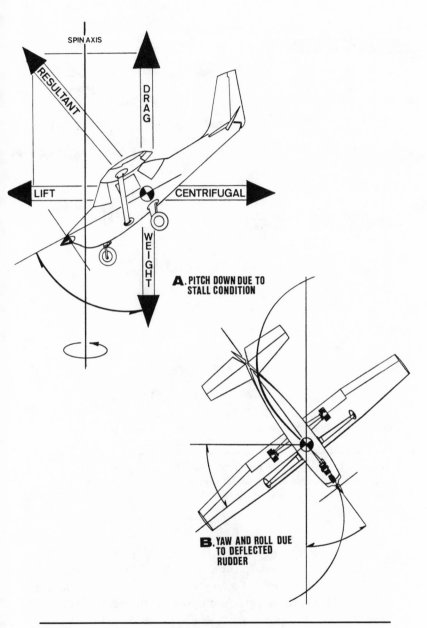

SPIN AXIS

RESULTANT

DRAG

LIFT

CENTRIFUGAL

WEIGHT

A. PITCH DOWN DUE TO STALL CONDITION

B. YAW AND ROLL DUE TO DEFLECTED RUDDER

FIG. 4.10. Spin entry includes (A) pitch down combined with (B) yaw and roll.

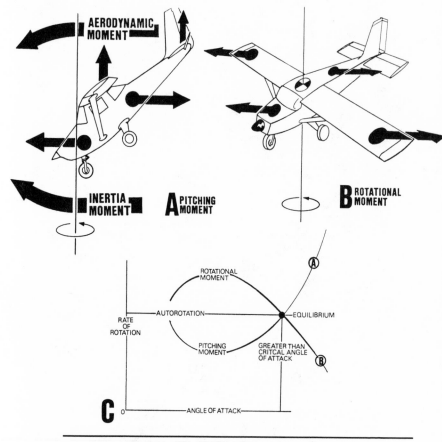

FIG. 4.11. (A) Pitching moments and (B) rotational moments are in equilibrium when the airplane is in spin autorotation.

P-factor is created because the descending propeller blade (as viewed from the cockpit) on the right side of the propeller disc has a greater angle of attack than the ascending blade on the left. This produces a greater thrust from the right side of the propeller disc that causes a left yaw reaction.

Changes in pitch produce different precessional reactions,

the amount depending on the rate of movement about the pitch axis and on the amount of engine power used. When the airplane is pitched down, there is a left yaw motion, and when the airplane is pitched up, there is a right yaw motion.

The effects of precession on a spin depend on such variables as rotation rate, angle of attack, and roll rate in combination with yaw. Although the size and weight of the propeller relative to the size of the airplane influence the effects of precession on a spin, the pilot can control this through the use of engine power.

The application of power in a right spin increases the up pitch. This induces a right yaw which, combined with the roll rate, increases the rate of rotation. When power is applied in a left spin, the spin tends to flatten because the precessional forces increase the pitch. The effects of precession on spins are general since the design of an airplane can significantly affect precession. However, with the power setting at idle, precession is minimized and recovery is more rapid. It is for this reason that idle power is used for spin recovery.

Cross-Control Spin Entry

A spin entry has been described from a coordinated control stall followed by rudder-induced yaw. While this is generally the procedure for teaching spin entry, most unintentional spins are entered from a cross-control situation. Two types of cross-control stalls, the slipping and the skidding types, can result in a spin entry either over-the-top or under-the-bottom, respectively.

Consider the skidding cross-control spin entry (Fig. 4.12). While in a left turn the ailerons are positioned opposite the turn, hence, the aileron is down on the inside wing and up on the outside wing. Simultaneously with aileron input, the left rudder is positioned to the inside of the turn. The airplane is now in a skid.

A comparison of the lift curves for the wings is presented in Fig. 4.13. The down aileron on the inside wing acts as a high lift

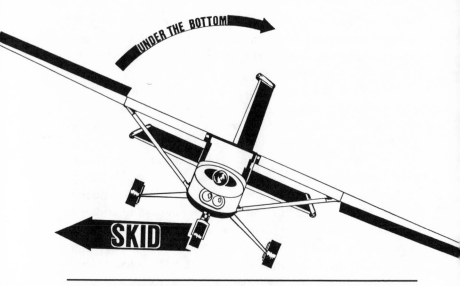

FIG. 4.12. Airplane in a cross-control skid.

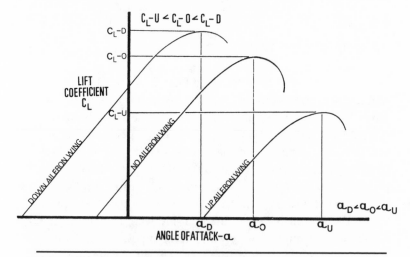

FIG. 4.13. Comparison of lift and angle of attack for wing conditions in a cross-control (skidding) turn leading to an under-the-bottom spin entry.

device (flap) thereby changing the camber of the wing. Consequently, the lift curve for this wing has changed. The wing lift has increased and the critical angle of attack has decreased. Conversely, for the outside wing, the lift has decreased and the critical angle of attack has increased. In addition, because of the skidding turn, the velocity of the outside wing is greater than that of the inside wing. If the nose is now pitched up, the critical angle of attack will be reached for the inside wing before that of the outside wing. The inside wing now stalls before the outside wing, and because of the lift differential between the two wings, the airplane rolls to the left (under-the-bottom). The increased drag on the inside (left) wing coupled with the yaw from the left rudder results in rotation to the left.

In the case where the controls are crossed in the opposite direction, the airplane is slipping (Fig. 4.14). Therefore, the outside wing stalls first because of the effects of the down aileron on the outside wing. The resulting lift differential rolls the airplane to the right (over-the-top), and the yaw produced by right rudder gives rotation to the right.

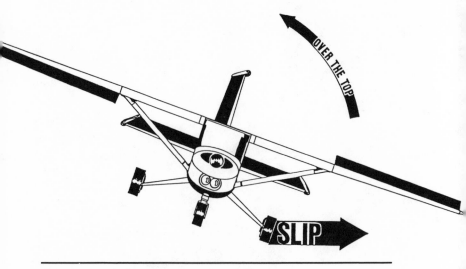

FIG. 4.14. Airplane in a cross-control slip.

AERODYNAMICS OF A SPIN

36

Summary

The spin is a complex aerodynamic maneuver. To describe it is not to understand it. To understand it, the pilot must be knowledgeable about aerodynamics. With this knowledge a pilot is better prepared to enter and recover from a spin. Just as important, this knowledge enables a pilot to avoid entry into a spin.

5
Airplane Design and the Spin

Background

Design technology associated with spins has received little attention, probably because the spin is not a normal mode of operation for most airplanes. The pilot usually lacks an understanding of the basic design principles associated with spins for two reasons. First, most general aviation airplanes are not required to recover from a developed spin, and, second, spin training is not required for pilot certification. However, to be spin proficient a pilot must be knowledgeable regarding design principles associated with the spin.

Design Factors

In general, there are three significant design factors associated with the spin: mass distribution, relative density, and tail configuration.

MASS DISTRIBUTION. The distribution of mass between the wing and fuselage determines how the airplane responds to control input. Mass distribution about the three axes and the rotation about these axes results in moments of inertia about all three axes. Because of the motion through the air, there are also aerodynamic moments acting on the airplane. In the developed spin these inertial and aerodynamic moments are in equilibrium. In order to recover from a spin, this equilibrium must be broken through control-induced aerodynamic moments.

Mass distribution can be configured in the wing or in the fuselage so that either one part dominates or the two are approximately equal. When mass is distributed primarily in the wing the moment of inertia in roll is greater than in pitch. Such distribution is found in airplanes with wing-mounted engines, fuel tip tanks, and high-aspect ratio wings. The elevator in the down position is the primary control for recovery.

A single-engine, tandem-seat airplane with the fuel tank located in the fuselage has a mass dominated fuselage. Here the moment of inertia is greater in pitch than in roll. The aileron, deflected into the spin, is the primary control for recovery. With the mass equally distributed between the wing and fuselage, there is a zero mass pattern. The moments of inertia in roll and pitch are approximately equal. The rudder is the primary control for spin recovery. It is deflected opposite the direction of rotation. Most single-engine, general-aviation airplanes have a zero mass distribution pattern.

RELATIVE DENSITY. A general factor associated with spins is relative density. This is a comparison of the airplane density with the density of the air in which it is flying. This is indicated by the relative density factor (RDF) and can have significant influence on effectiveness of control input on spin recovery. The numerical value of RDF is calculated as follows

$$RDF = m/\varrho Sb$$

where m = mass, ϱ = air density, S = surface area, and b = wing span. The RDF varies directly with weight and inversely with altitude. Gross weight changes from fuel, cargo, and passengers do not generally cause a significant effect on relative density for most airplanes, but weight changes associated with structural alterations combined with a wide range of operational altitudes does have the potential for relative density effects. High RDF airplanes require more rudder and elevator effectiveness for spin recovery as compared to those with low RDF.

TAIL CONFIGURATION. The configuration of the tail has considerable importance in the spin recovery, particularly with a zero mass distribution airplane. Since the rudder is the primary recovery control is usually in the stalled wake of the horizontal tail and possibly the wing (an area where the dynamic pressure is low), design considerations have a profound influence on spin recovery. In order to be effective, in other words to change moments, a substantial part of the rudder must be outside the stalled wake area. The fixed area under the horizontal tail must be sufficient to damp the spin motion so the flight path steepens and the rotation slows. The portion of the rudder in the stalled wake of the horizontal tail depends on the position of the horizontal tail. The shielding effects of different horizontal tail positions on the rudder are shown in Fig. 5.1.

Rudder effectiveness depends not only on position of the horizontal tail but also on the length of the rudder relative to the horizontal tail. A full-length rudder extending below the horizontal tail provides additional unshielded rudder as compared to a partial length. With the sweptback vertical tail it is possible to have a large portion of both the vertical tail and rudder shielded.

The fuselage area and unshielded rudder area have been considered important factors in spin characteristics and recovery. Two mathematical factors, body damping ratio (BDF) and unshielded rudder volume coefficient (URVC), have been used

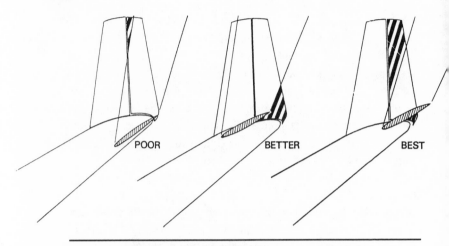

FIG. 5.1. Shielding effects of different horizontal tail positions on the rudder.

to formulate a spin damping power factor (DPF). This formulation has been modified to be the tail damping power factor (TDPF). Although this mathematical formulation has been used for prediction of spin recovery effectiveness, recent research makes its use of questionable value simply because other variables, such as control position, wing position, fuselage shape, not accounted for in the formulation, give different spin characteristics for a given tail design. However, this does not negate the importance of tail configuration in the recovery from a spin.

NASA Model Airplane Spin Testing

NASA Langley Research Center has conducted an extensive stall/spin research program that includes wind tunnel tests, radio-controlled model tests, and full-scale flight tests. Wind tunnel spin testing was conducted with a model of a light, low-wing, single-engine, general-aviation airplane to determine the effects of tail configuration on spin characteristics (Fig. 5.2).

FIG. 5.2. Model airplane being tested in NASA's wind tunnel. Courtesy of NASA.

NASA determined that tail configuration can significantly influence spin and recovery characteristics, but prediction of spin characteristics cannot be based solely on tail design. For example, aileron deflection had an adverse effect on spin recovery characteristics. Therefore, tail design effectiveness based only on aileron neutral position does not correctly predict spin/recovery characteristics.

Spin tunnel tests on the model airplane did show that the T-tail and the tail configuration with the horizontal tail mounted halfway up the vertical tail produced the best spin recovery (Fig. 5.3). There were, however, other geometric design features that affected spin recovery characteristics.

A sharp angled fuselage bottom resulted in flat spins, whereas, rounding the fuselage bottom eliminated flat spins. A

42

ventral fin was also found to be effective in eliminating a flat spin. However, both a round fuselage bottom and a ventral fin located under the fuselage-wing junction improved spin recovery characteristics, but the fin placed under the tail had no appreciable effects (Fig. 5.4).

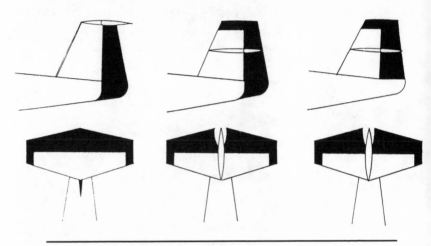

FIG. 5.3. Tail configurations found by NASA to give good spin recovery characteristics when used on a model airplane.

The junction of the wing trailing edge with the fuselage produced an airflow characteristic that affects spins. The airflow was so sensitive that a sharp-edged wing fillet produced flat spins, but this was eliminated by rounding the edges of the fillet. It appeared that the sharp-edged wing fillet disturbed the airflow over the tail.

The simultaneous application of elevator reversal with rudder reversal did not consistently improve recovery and, in some cases, actually delayed recovery. However, the use of the elevator alone did not stop a developed spin.

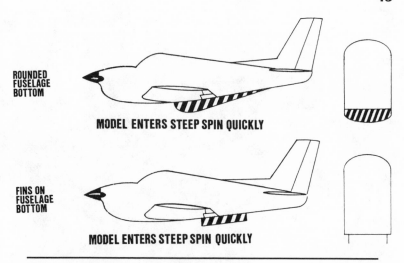

FIG. 5.4. Fuselage modification and strakes
found to improve spin recovery
characteristics of a model airplane.

NASA Airplane Spin Testing

In order to better understand the effects of configuration
changes on spin and recovery characteristics, NASA conducted
extensive tests with a low-wing, single-engine, general-aviation
airplane. The airplane used was the Grumman American AA–1
Yankee (Fig. 5.5). The parameters investigated were tail configuration, fuselage modification, moment-of-inertia variations,
center-of-gravity (CG) positions, and control inputs.

A total of 529 spins (3559 turns) were investigated. All
spins were entered with the wings level and power at idle. The
airplane spin recovery parachute was required 16 times.

In the NASA tests the horizontal tail position and the rudder lengths were changed to duplicate those used in wind tunnel
spin tests of the model airplane. The spin modes and recovery

FIG. 5.5. Grumman American AA–1 Yankee used by NASA to study configuration changes on spin characteristics. Courtesy of NASA.

characteristics agreed fairly closely with test results on the model airplane. However, regardless of the tail configuration, normal control inputs (rudder against spin rotation followed by down elevator) provided the quickest recovery. While simultaneous control input was almost as good as the normal input, simply neutralizing the control produced a slower recovery. When recovery was attempted with rudder reversal alone, the airplane would recover from left spins but not right spins for three of the tail configurations.

Moment-of-inertia variations were produced by addition of ballast while still keeping the CG constant. Basically, addition of mass to the wings slowed the recovery; however, normal recovery procedures were the least affected. If there was asymmetry in mass distribution (one wing heavier than the other), a flatter spin resulted that required nearly three times as many turns for recovery. Rudder reversal alone and neutralizing controls did not give recovery with a wing-heavy condition. One tail configuration, when combined with asymmetrical wing loading, produced a spin that required spin chute deployment for recovery. This configuration consisted of a short rudder with a low, aft, horizontal tail.

Movement of CG aft had little effect on spin and recovery characteristics for the various tail configurations. As the CG moved aft the spin rate decreased slightly and recovery was slightly faster. The CG was kept within limits for all tests.

A drooped wing was created by the addition of more camber at the leading edge (Fig. 5.6). This drooped wing caused the

FIG. 5.6. Drooped wing created by addition of more camber to the leading edge. Courtesy of NASA.

airplane to spin flat for tail configurations that did not previously have flat spin modes. This flat spin mode did not require any special control inputs. Wind tunnel spin tests did not predict this mode.

Various combinations of strakes were located at different fore-aft and lateral positions on the underside of the fuselage. These modifications were tested with a tail configuration consisting of short rudder and a low, rearward-placed horizontal tail. None of the fuselage modifications eliminated the flat spin mode. A ventral fin did reduce angle of attack and slowed rotation rate, but the spin was still moderately flat.

In conclusion, it should be noted that for all the configurations tested the airplane consistently recovered from one-turn spins within one turn following normal recovery input; however, some configurations had unrecoverable spin modes after 3 to 6 turns.

Wing Design

Because wing design had such a prominent influence on spin characteristics, NASA conducted more extensive tests on wing leading-edge modifications. One type of wing design delayed wing-tip stall and increased both stall departure and spin resistance of the airplane (Fig. 5.7). Basically the wing modification was an abrupt planform discontinuity between the basic wing and the outer panel. The partial span discontinuity consisted of an outboard leading-edge droop that increased the camber of the wing. Although additional flight tests of this design parameter are needed, the discontinuous leading edge of the wing appears to be a significant factor for increasing spin resistance.

DROOPED LEADING EDGE

BASIC LEADING EDGE

FIG. 5.7. Wing modification used to study spin resistance. Courtesy of NASA.

Other Geometric Design Factors

There are other geometric design features that have been found to influence spin characteristics and recovery.

With all other factors equal the high wing airplane has better spin recovery due to the higher dihedral and airflow wake characteristics in the vicinity of the tail. The wake appears to pass over the tail. In general, flap extension flattens the spin and slows the spin rate, but this is less pronounced in a high-wing airplane. For a low-wing airplane, flap extension reduces the air-flow over the tail and thereby reduces rudder effectiveness in recovery.

While structural mass of the airplane is fixed in flight, the shift in fuel quantity results in different mass distribution, particularly since fuel will shift during rotation. Therefore, location of the fuel tanks becomes important when an airplane spins. This is apparent when fuel tanks are in the wing-tip area and even along the wing spar. As weight increases along the wing, the effectiveness of the elevator as the primary recovery control increases. Hence spin recovery technique depends on the fuel load configuration at the time of the spin.

Spin recovery depends on control effectiveness. Rudder and elevator effectiveness improves at lower angles of attack and decreases at high angles of attack. Consequently, any design feature that promotes flat spins also decreases control recovery effectiveness. For example, the distance from the CG to the tail influences angle of attack and rate of rotation in the spin. A short length produces flat spins at a slow rate whereas increasing the length gives low angles of attack and higher rates of rotation. This is the reason an aft CG flattens a spin and makes recovery more difficult. The distance from the CG to the tail is the moment arm. As this distance decreases the aerodynamic moments become less and there is insufficient control effectiveness for recovery. The lowering of flaps flattens a spin and reduces rotation rate, probably due to air wake disturbance over the tail.

Summary

The influence of design factors on spin recovery characteristics is difficult to evaluate because the factors are interrelated. The geometric design features presented here are not intended to be inclusive but are presented to illustrate the complexity of the problem. Most important, this sample of design factors illustrates the need to adhere to manufacturer recommendations for spinning an airplane. However, a pilot must realize that these recommendations are based on spin tests conducted within the guidelines of certificate requirements and may not reflect all combinations of variables.

6

Recovery from the Spin

Historical Background

The early days of aviation were a dangerous age of trial and error where even the smallest mistake was usually fatal. This did not deter persons from wanting to be pilots or pilots from testing new airplane designs. One such test pilot was Wilfred Parke. During one particular test flight of a Roe Avro monoplane in 1912, Parke put too much back pressure on the control stick while the airplane was in a spiral glide. Suddenly the airplane dropped into a whirling spiral descent. The airplane had entered what was then called a spiral dive, now called a stall/spin. Few pilots had recovered from this flight attitude, and those who had did not know how they had achieved the recovery.

Parke changed throttle settings and control stick positions, but the spiral descent continued. He put the rudder into the spin but nothing happened. Parke then put the rudder opposite the spin direction and the spinning stopped. He pulled the airplane out of the resulting dive. Hence, the "Parke's dive" was born,

50

and the rudder technique for recovery became standard procedure.

Less than four months later Parke was testing a Handley Page monoplane. On takeoff the engine failed. He attempted to turn back to the field, stalled, and spun in. It seems that Parke had discovered the recovery technique for a spin, but he did not totally comprehend the factors leading to the spin. As in the case with most stall/spin accidents, he died.

The spin became an aerial combat maneuver during World War I. However, spin recovery remained a more or less hit-and-miss procedure. It was not until the 1920s that a consistently applicable theory was developed as a result of the research work of Frank W. Gooden. Over the years spin recovery procedures have become more comprehensive as research has provided better understanding of the factors involved in spins.

Normal Spin Recovery

For most single-engine, general-aviation airplanes a standard or basic procedure for spin recovery has evolved. The key factors are as follows:

1. Determine the direction of rotation
2. Neutralize the ailerons
3. Close the throttle
4. Apply full rudder opposite to the direction of rotation
5. Brief pause
6. Move the control yoke (stick) briskly forward, full forward if required
7. Hold control input until rotation stops
8. Neutralize rudder when rotation stops and carefully recover from the resulting dive

Although these steps are used as the "standard" recovery procedure for single engine general aviation airplanes, it must be noted that the recovery procedure may differ for individual airplanes.

Spin versus Spiral

A common problem in the recovery is pilot confusion as to the direction of rotation, a critical determination since recovery depends on correct application of the rudder. The turn indicator indicates the direction of rotation regardless of location on the instrument panel. However, the turn indicator does not differentiate between a spin and a spiral. While the spin involves autorotation, the spiral is a diving turn with a low angle of attack. Therefore, the turn indicator shows the direction of turn for both a spin and a spiral. The indicated airspeed differentiates a spin from a spiral. The indicated airspeed during a spin is stabilized, whereas the indicated airspeed is increasing in a spiral. This differentiation is critical since the recovery procedures are different for spins and spirals (Figs. 6.1 and 6.2).

Recovery from a spiral requires coordinated control pressures. The wings are leveled by coordinated aileron and rudder pressures followed by back pressure to bring the nose up. Recovery from a spin is a mechanical step process whereby the pilot responds with control movements rather than control pressures. Consequently, correct differential between the spin and spiral is imperative if proper recovery procedures are to be initiated.

Control Input Sequence

Application of rudder opposite to the spin direction produces an aerodynamic moment that opposes spin rotation (Fig. 6.3). This yawing moment reduces the rate of rotation and the angle of attack. Forward yoke movement produces a nose-down aerodynamic moment that reduces the angle of attack (Fig. 6.4).

FIG. 6.1. Airspeed indicates a spin and turn indicator shows right rotation. (Photo by author)

FIG. 6.2. Airspeed indicates a spiral and turn indicator shows left rotation. (Photo by author)

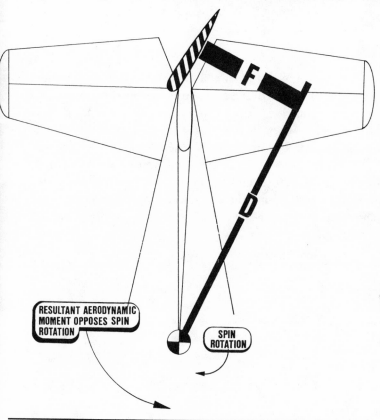

FIG. 6.3.   Rudder deflection produces an
aerodynamic moment that opposes spin
rotation.

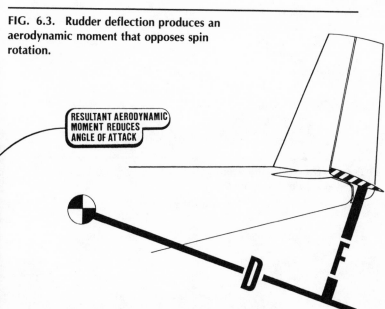

FIG. 6.4.   Down elevator produces an
aerodynamic moment that reduces angle of
attack.

The sequence of rudder followed by elevator movement is important. The effectiveness of the rudder in slowing rotation and reducing angle of attack depends on the unshielded portion of the rudder. When the elevator is up a larger portion of the rudder is usually exposed so that as the rate of rotation is slowed by the rudder, the elevator is becoming more effective (Fig. 6.5). Down-elevator before rudder application reduces the unshielded area of the rudder, and the effectiveness of the rudder in stopping rotation has been significantly reduced, perhaps to a point of no recovery.

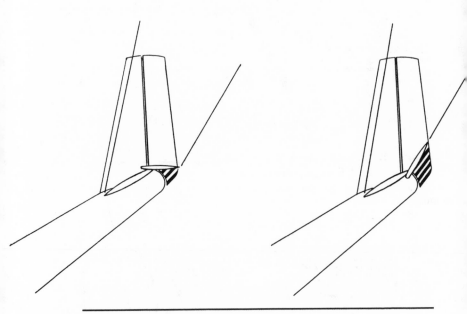

FIG. 6.5. Down elevator reduces the unshielded area of the rudder.

Summary

It is important to note that while research effort has been directed at determining a standard spin recovery procedure, the combination of design factors and airplane configurations produce different recovery characteristics. For example, in some airplanes the yoke (stick) must be moved to the neutral position while in others the yoke must be moved forward as soon as full opposite rudder is applied. Again, a pilot must be familiar with the procedures given in the airplane operation handbook and must abide by the steps given for spin recovery.

7
NASA
Spin
Testing

Historical Background

During the 1930s and the 1940s, the National Advisory Committee for Aeronautics (NACA) studied flight characteristics of airplanes designed and built during this era. Because of military needs, research efforts were directed toward aircraft performance needed in a combat environment. During the 1950s and 1960s, research attention was directed to the rapid change in military aircraft designs brought about by jet propulsion. In the early 1970s there was a research program initiated for purpose of developing new stall/spin technology for general aviation aircraft. Consequently, a general aviation stall/spin program became a major research project at the NASA Langley Research Center. The program included a large spectrum of analytical studies involving wind tunnel testing, radio-controlled model testing, and testing of light, single-engine airplanes. Of interest to the general aviation pilot is the results of the stall/spin tests for two aircraft types, the single-engine, high-wing, conventional-tail, light airplane and the single-engine, low-wing, T-tail, light airplane.

Single-Engine, High-Wing, Conventional-Tail Airplane

This particular airplane (Fig. 7.1) was selected for study because the production airplane had a relatively good stall/spin accident record as compared to other similar type airplanes. All spin tests were conducted in the utility category as recommended by the manufacturer, although certain test maneuvers were either not approved or specifically prohibited. The test program consisted of 128 spin maneuvers performed in an unmodified

FIG. 7.1. Single-engine, high wing,
conventional-tail airplane tested by NASA.
Courtesy of NASA.

airplane. A summary of the results is presented in the following sections.

STALL. The stall was investigated since it is the flight condition usually needed for spin entry. Also, the stall was investigated to see if the airplane would enter a spin with no rudder input. With idle power and a slow approach to minimum airspeed the airplane reached a maximum angle of attack of 16 degrees, but no stall developed since 18 degrees was the critical angle of attack. A higher angle of attack could not be obtained due to insufficient up-elevator caused by flexibility in the control system. However, with a maximum power setting, an angle of attack of 20 degrees was obtained, which was sufficient to produce a stall. The added power gave additional upward pitch moment and increased the dynamic pressure on the tail. The stall break was described as a gentle pitch down with a right roll of 40 degrees, a right wing drop, and a heading change of 70 degrees.

SPIN. It was difficult to compare characteristics of six-turn left and right spins because spins to the right would transition into spirals before three turns. This was attributed to elevator control system flexibility. Despite the pilot holding full aft control input the elevator control surface became less negative until it was 10 degrees less than the full-up position. The lack of full-up elevator reduced the nose-up pitch of the airplane sufficiently so as to allow transition into a spiral.

The left spin remained in the spin mode for the full six turns. The angular velocity stabilized at 200 degrees per second. The angle of attack gradually decreased but remained approximately 2 degrees above stall. This decrease was probably due to a gradual decrease in up-elevator (7 degrees) as the spin progressed, but the decrease was less than the 10 degrees found in the right spin. There was an abrupt decrease in angle of attack with recovery control input. A right sideslip of 20 degrees existed throughout the spin.

While the differences in right and left spins were attributed to elevator deflection differences, it was noted that right and left rudder travel also differed. In fact, the pilot was able to obtain different rudder deflections simply by varying pressure on the rudder pedal.

Since control flexibility seemed to influence spin characteristics, tests were conducted with the elevator control cables tightened to nearly eight times the normal required tension. With the controls tightened the elevators maintained a greater up position, but the resulting spin modes were basically the same as those with the more flexible controls.

ENGINE POWER. With idle power it was not possible to create an angle of attack high enough to stall the wing, hence no spin was possible. With full power an upward pitching moment sufficient to stall the wing was generated. Although a spin to the left was possible with partial power, a spin to the right was not possible unless entered with full power. It was concluded that the aerodynamic upward pitching moment produced by power was more important than the downward pitching moment produced by the propeller gyroscopic action.

AILERON INPUT. Aileron input affected the characteristics of the spin. Aileron input was delayed approximately one second after spin entry; however, when aileron was applied it did not appreciably alter the results. In general, aileron away from the spin slowed the rotation and flattened the spin whereas aileron into the spin increased the rotation and decreased the angle of attack.

Aileron input against the spin produced an angular velocity of 75 degrees per second as compared to 200 degrees per second with the aileron neutral. The angle of attack was approximately 26 degrees but dropped abruptly to about 13 degrees after one turn. The angle of sideslip increased to well over twenty degrees. Ailerons into the spin produced an angular velocity of 250 degrees per second. The angle of attack decreased to 16 degrees, less than the 18 degrees required for stall, but the angle of attack for the wing with the up-aileron, the retreating wing, was above the required stall angle. The sideslip decreased to zero. A strong oscillatory motion was noted. These oscillations had the potential to overload the airplane.

CONTROL INPUT REQUIRED FOR RECOVERY. In general, airplane response was prompt when any recovery control input was made. Control input slowed the spin rate and decreased the angle of attack.

Application of opposite rudder and nose-down elevator stopped the spin in one-half turn. Release on any *one* of the prospin controls resulted in a decreased angle of attack and decreased spin rate. If all prospin controls were released the angle of attack immediately decreased since the elevator moved to a negative 5-degree position. The spin rate momentarily increased because the aileron deflected in a prospin direction. The angle of attack and spin rate then decreased as the aileron moved to the neutral position. The airplane usually recovered within one-half turn after control input was applied.

SUMMARY. The spin characteristics of a production model general aviation airplane were documented, and while the results were not consistently repeatable, the findings have practical importance for the general aviation pilot:

1. The recovery response met the certification requirements of FAR 23 for a utility category airplane; however, the airplane would transition from a spin to a spiral, a transition that requires pilot recognition in order to execute the proper recovery.
2. Although the recovery met certification requirements, the characteristics of the spin varied simply because of the many variables associated with an aerodynamic spin; therefore, a pilot can expect different spin modes for the same model airplane. The flexibility of the control system is but one example of how one variable affects the

characteristics of a spin. With this flexibility reduced and a slightly different control input (one-half aileron into spin) at entry, the spin changed modes at three turns and two turns rather than the usual one-half turn were required for recovery.

3. Strict adherence to the airplane placard and operation manual is imperative. For example, when the spin was performed with the flaps extended, a procedure specifically prohibited for the airplane, the airspeed buildup was so rapid the test was terminated to prevent damage to the airframe. Also, aileron input influenced the spin characteristics, supporting the recommendation that the aileron be neutral to recover from a spin.

Single-Engine, Low-Wing, T-Tail, Light Airplane

The test airplane was a four-place, single-engine, low-wing, retractable gear, T-tail design with a stabilator. Although the airplane was a one-of-a-kind research aircraft (Fig. 7.2), it was

---

FIG. 7.2. Single-engine, low-wing, T-tail airplane tested by NASA. Courtesy of NASA.

considered to be representative of this type of general aviation airplane. Modifications had been made to the wings. Starting with the taper portion of the wing the leading edge drooped. All tests were conducted at gross weight and at aft center-of-gravity positions.

STALLS. Approach to a stall was indicated by a light buffet that progressed to a heavy buffet at the stall break. There was a slight roll-off tendency following stall when a zero sideslip was maintained; however, the roll-off became uncontrollable when sideslip was present at the stall. The airplane stalled at about a 20-degree angle of attack.

A stall with idle power produced a roll motion, but the roll motion became more uncontrollable with up-stabilator deflection. When stalled in a sideslip the airplane rolled away from the sideslip at stall. However, stalls from coordinated 30-degree and 60-degree banked turns produced minimal roll-off.

When stalled at maximum power the airplane usually rolled right when in zero sideslip. Stalls at full power with 30-degree and 60-degree bank turns resulted in roll-offs to the right. With sideslip at stall the airplane rolled away from the slip and could not be controlled until the airplane was unstalled. With 40 degrees of flap deflection the airplane rolled to the right regardless of power setting.

SPIN. Of the 209 spin attempts, 173 resulted in a spin, 34 in a spiral, and 2 failed to produce either a spin or a spiral. Spin characteristics for different aileron positions are described in this section.

When the ailerons were in a neutral position the airplane spun at a 43 degree angle of attack with a rotation rate of 2.7 seconds per turn while descending at 120 feet per second. The angle of attack was not affected by deflected ailerons, extended landing gear, deflected flaps, or power setting. The spin was oscillatory in roll and pitch, the rate varying over a period of 4 to 4.5 seconds, but roll and pitch oscillations were not in phase. The airplane spun with a slipping rotation.

Deflecting the aileron with the spin at entry tended to make the airplane spiral rather than spin, and it also made the differentiation between spin and spiral difficult. Aileron against the spin at entry tended to stop the spin entry, but this deflection did not stop the spin after the initial quarter turn. Aileron deflection with the spin increased the magnitude of oscillation in roll and pitch, whereas deflection against the spin reduced the oscillation. Ailerons with the spin resulted in a slipping rotation whereas ailerons against the spin resulted in a skidding rotation. Aileron position did not change the spin angle of attack. Aileron into the spin increased the roll and pitch oscillations, but aileron away from the spin reduced the magnitude of the oscillation.

SPIN RECOVERY. While different antispin control inputs were investigated, normal recovery input gave the fastest and most consistent recovery from steady spins with the power at idle and the airplane in clean configuration. Normal recovery was application of the rudder against the spin prior to down deflection of the stabilator. However, the initial application of these controls did not always stop the spin.

Neither neutralizing the controls, releasing the controls, nor deflecting the stabilator stopped a developed spin. Rudder deflection against the spin as a single control input did not always stop a spin and, when combined with aileron against the spin, would not stop any spin mode.

All one-turn spins recovered within one and one-fourth turns after recovery control input. Spins of more than three turns required more turns and longer times for recovery. Generally, timing of control inputs was a key factor. Application of rudder prior to moving the stabilator resulted in a faster spin recovery.

SUMMARY. The airplane used in this investigation was a unique research airplane. Consequently, the results cannot be extrapolated to include an unmodified production airplane. From a pilot's perspective, however, the results have practical implications.

    1. It is important that a pilot follow manufacturer recommendations for spin recovery procedures. Not only is the correction control input important, but the sequence and timing of such controls are just as essential.

    2. Again, just as was noted in the results of the high-wing airplane spin tests, it is difficult to determine whether the airplane is entering a spin or a spiral, a critical factor since this differentiation determines the correct control input.

Summary

Although the spin test results discussed here represent but two general aviation airplanes, these results have significant meaning for all pilots. First, the results were documented by elaborate instrumentation. Consequently, the results are objective rather than subjective evaluations based on pilot opinion. Second, although the airplanes satisfy the spin certification requirements of FAR 23, there is considerable variance in spin characteristics simply because of the many variables associated with the aerodynamic spin. It can be concluded that any pilot, regardless of expertise or experience, should treat the aerodynamic spinning of any airplane as a venture into the unknown. The spin characteristics may be docile and predictable for a given set of conditions, but the deviation in one variable may be the factor needed to produce an unexpected spin mode.

8
The Incipient Spin

Background

A pilot is turning from a left base leg to final approach with a quartering left crosswind. Because of the wind, the turn rate is such that the airplane will not be aligned with the runway but will instead overshoot final approach. Consequently, the pilot banks the airplane more steeply by using more bottom rudder in order to turn the nose to the left and simultaneously maintains proper pitch by applying back pressure to increase the angle of attack. Suddenly, the low wing drops sharply and the airplane rolls to the left. The airplane has entered a spin "under the bottom" from a cross-controlled stall. This type of stall/spin is usually fatal since it is at an altitude of 1000 above ground level (AGL) or less.

This example was cited to illustrate that while a pilot may be proficient in recovering from a multiturn spin and the airplane may have excellent multiturn spin recovery characteristics, these may be of little value in the recovery from an incipient spin. The characteristics of a developed multiturn spin are not necessarily the same as those of an incipient spin, and, because of the close proximity to the ground, the pilot may not initiate the proper control input for recovery.

The Incipient Spin

By definition, the incipient phase of a spin is from the time the airplane stalls and rotation begins until the spin axis becomes near vertical. During the transition from horizontal flight path to a vertical path, the rotation rate is increasing. The airplane may be in the incipient phase two to three turns before entering the developed phase.

From the pilot's viewpoint, entry into an incipient spin begins with a roll during the stall. As the airplane pitches down the yaw motion begins. By the half-turn point the pitch is vertical, but the flight path is inclined. At the one-turn point the flight path is now vertical and the pitch is increasing. As the rotation continues into the second and third turns the rate of rotation increases and the pitch, yaw, and roll motions follow an oscillatory pattern.

From the aerodynamic viewpoint, the incipient spin is a transition from a stall condition to a steady, developed spin. Pitch and yaw couple to produce a rolling moment about the longitudinal axis of the airplane. While the roll rate is increasing there is an increase in oscillation. This reflects an imbalance between aerodynamic forces and inertia forces. Pitch oscillation is approximately 180 degrees out of phase with roll oscillation. The developed phase (autorotation) begins when aerodynamic and inertia moments are in equilibrium.

Incipient Spin Tests by NASA

While the incipient spin characteristics of general aviation airplanes are the least documented, the NASA spin research program has included, as part of several spin investigations, the study of the incipient spin. The incipient spin was included in the spin testing of both high- and low-wing general aviation airplanes. In both cases the incipient testing was confined to one turn.

HIGH-WING AIRPLANE. For the high-wing airplane with a conventional tail the angular velocity was 120 degrees per second at the one-turn position, and the angle of attack oscillated around the stall value of 18 degrees. Recovery at the one-turn position required less than a one-half turn and 1 second. While spin motion either right or left was oscillatory, spins to the left exhibited greater magnitudes of pitch, roll, yaw, and sideslip. Although there was resistance to spin entry to the left when turning right, the airplane entered a left spin from a left turn quite readily.

LOW-WING AIRPLANE. The incipient spin for the low-wing, T-tail airplane was described as rolling motion superimposed on pitching motion. During this transition phase it was difficult to determine if the airplane was entering a spin or a spiral. During the actual spin the altitude loss averaged 120 feet per turn. Altitude losses during recovery ranged from 200 to 600 feet with an average of 400 feet. The recovery time was 1.1 to 4.0 seconds for an average of 2.5 seconds. The dive following recovery produced an altitude loss of 300 to 1200 feet, averaging 600 feet. The time required ranged from 1.9 to 6.5 seconds for an average of 4.1 seconds. Recovery to level flight was limited by the maximum load factor of 3.8 G.

As a whole, from stall to pullout at level flight, a one-turn spin required 300 to 2100 feet of altitude and from 7.4 to 19.8 seconds. The average was 1160 feet of altitude loss and 12.8

seconds. Thus, a stall going into a one-turn incipient spin at airport traffic pattern altitude of 800 to 1000 feet AGL would require immediate recognition and reaction by the pilot, something that would be unlikely since it is difficult to differentiate between a spin and a spiral at this point in the spin.

In another NASA study a modified low-wing, single-engine, general-aviation airplane with a stabilator-type tail was used to specifically investigate incipient spin characteristics. The basic airframe had been modified with antispin fillets and strakes. This was an optional modification supplied by the airplane manufacturer. This modification permitted intentional spins within the specified weight and balance limitations.

All spins were entered from a stall at level flight with idle power. From the stall, spin entry was a smooth, rolling motion produced by the coupling of roll and yaw. The pitch decreased rapidly to a 50- to 55-degree, nose-down attitude at the one-turn position.

All spin recoveries were initiated either at the one-half or one-turn positions. For a one-turn spin, full aileron opposite the spin direction was applied as the airplane rolled through a 90-degree bank. Aileron input was not used in the one-half turn spin. However, regardless of the control sequence, the stabilator was the primary recovery control. Returning the stabilator to neutral produced an immediate decrease in pitch and angle of attack, accompanied by a significant increase in roll rate. Recovery took place within 1 second, during which time the airplane rotated approximately one-fourth turn. The only way a pilot could visually distinguish a spin from a spiral was by reference to the airspeed indicator.

Summary

When interpreting this research data on the incipient spin, one must remember that if such variables as mass distribution, center of gravity, control input sequence, or entry conditions were changed, the spin and recovery characteristics may differ considerably from those presented. There are, however, several factors common to the three NASA investigations that must be noted. First, an airplane can transition from a spin to a spiral and it is difficult for the pilot to distinguish one from the other. Second, an incipient spin at low altitude requires immediate differentiation between a spin and spiral followed by correct control input. The airspeed indicator provides the pilot a means to differentiate between a stall and spin, but at low altitudes adequate time may not be available. The key lies with the pilot's ability to both correctly recognize and react.

9
Spinning
a
Cessna 150

Background

In late 1958 the Cessna Aircraft Company introduced the first Cessna 150, a two-place trainer that was to continue in production until 1977. This airplane has been called the ideal trainer, and while this may be debated, the statistics are impressive. For example, in 1969 there were nearly seven million flight training hours flown in the United States. Of this total the Cessna 150 accounted for over three million flight training hours. The safety record is equally impressive. The airplane has the lowest accident rate and, when crashed, has the highest survival rate of any two-place trainer.

As a utility category airplane the Cessna 150 is approved for intentional spins; however, the spin recovery procedure was not given in the airplane operational manual until 1972 (Model 150L). In 1976 the airplane operation manual (Model 150M) presented specific procedures for the pilot to check prior to intentional spins. Instructions were given for entry into and recovery from a spin. In 1980 the Cessna Aircraft Company published a booklet, *Spin Characteristics of Cessna Models 150, A150, 152, A152, 172, A172 & 177,* that contained basic guidelines for intentional spins. The booklet also contained brief descriptions of spin characteristics for recent Cessna models approved for spins.

Spin Exploration

For purposes of instructing pilots in the characteristics of a spin, the author investigated and documented the spin performance of two Cessna 150 airplanes (Model 150L and Model A150M) (Fig. 9.1). All spins were conducted with approximately mid-CG and weight approximately 1400 pounds. No special instrumentation was used other than a video camera for recording the instrument panel. Power at spin entry was either idle or 1700

**FIG. 9.1.   A Cessna Model 150 Aerobat.
Courtesy of Cessna.**

RPM. Normal control inputs as recommended by Cessna (power idle, opposite rudder followed by down elevator) were used, and recovery from the dive was confined to approximately a 3.0 G pull-out. The results were as follows.

ONE-TURN SPIN. When the airplane entered the spin with power idle, it appeared to be a steep turn, whereas with power the airplane appeared to roll over on its back. At the one-turn position the pitch was near vertical. If the power was reduced to idle at the one-fourth turn position the pitch oscillated, but if reduced to idle at the one-half position, pitch oscillation was minimal. With power at idle, recovery was within one-fourth turn for turns to the left, but to the right recovery was within one-eighth turn. Recovery was approximately one-fourth turn longer for left spins when power was used at entry; however, power did not affect recovery to the right.

Entry into the spin with power idle resulted in average altitude losses of 216 feet for one turn and 585 feet for recovery with an average total loss of 802 feet. Comparison of right and left spins showed the right spin to have slightly less altitude loss. The average time for a one-turn spin was 5.33 seconds—6.35 seconds for recovery and 12.2 seconds total.

With power at 1700 RPM the altitude loss was slightly greater. When power was reduced to idle at the one-fourth turn position, the average loss for the one turn, recovery, and total was 218 feet, 600 feet, and 812 feet as compared to 225 feet, 626 feet, and 853 feet when power was reduced at the one-half turn position. While the earlier reduction in power resulted in less altitude loss, the difference was not significant. As would be expected from the altitude losses, the times were nearly the same, and the times were only slightly longer (approximately 1 second) than the spin entry with idle power.

CROSS-CONTROL ENTRY. Spin entries were attempted for left and right turns with controls crossed to produce either spins over-the-top or under-the-bottom. For spins under-the-bottom, entries were consistent regardless of turn direction. If any one of the control inputs were relaxed during entry the airplane would transition to a spiral. The airplane would not enter a spin over-the-top with power idle and with power it would enter only from turns to the right if the power was 2000 RPM or more.

RECOVERY CONTROLS. Control inputs for recovery followed the normal recovery procedures recommended by Cessna. It was noted that altitude losses were dependent on the pilot's reaction. In fact, simultaneous application of recovery controls did not seem to influence recovery; however, any delay in application of either rudder or elevator did give greater altitude loss. All pull-ups from the recovery dive were confined to approximately 3.0 G and were within the "never-exceed-airspeed" limits.

There was a series of tests utilizing the no-hands spin recovery. This is a spin recovery technique developed by aerobatic pilot Eric Muller and explored by Gene Beggs. It has been advocated as an emergency recovery procedure to be used for accidental spins. The steps to be used in this recovery technique are as follows:

1. Reduce power to idle
2. Release the control yoke
3. Apply full rudder opposite direction of rotation
4. Neutralize rudder when rotation stops
5. Recover from dive

This technique was used to recover from right and left one-, two-, and three-turn spins. The average altitude losses are given in Table 9.1.

There was some difference in average altitude losses for spins in the two different models of aircraft, but there was only a slight difference when comparing right and left spins. Obviously, the altitude losses were greater as the number of turns increased. It was noted that recovery losses were dependent on how quickly the pilot resumed control of the yoke. This also

Table 9.1. Average Altitude Losses (Feet) for Spins in a Cessna A150M and a Cessna 150L with a No-Hands Recovery

	Spin	Recovery	Total
One Turn			
A150M	233	633	895
150L	230	515	855
Two turn			
A150	480	870	1350
150L	495	735	1230
Three Turn			
A150	740	930	1625
150L	825	860	1605

Source: author's research.

directly influenced the airspeed during the recovery from the dive. The airspeed approached red-line value when recovering from the three-turn spin. When the no-hands technique was compared with the normal technique for a one-turn spin, the altitude losses were essentially the same.

The movement of the control yoke during the recovery phase of a no-hands procedure followed a consistent pattern. When released the yoke moved forward to neutral, rotated into the spin, and then back to neutral after which it moved slightly aft. The movement of the control yoke into the spin and back to neutral was rapid and positive.

Although the no-hands technique would recover both Cessna models from spins up to three turns, after three turns the no-hands technique would not work. The yoke would move forward when released, turn in the direction of the spin, and remain in this position. The airplane would remain in the spin and required normal control input for recovery.

SUMMARY. The spin evaluation of the Cessna 150 was confined to the incipient phase of the spin with particular attention directed to the one-turn spin. It was concluded that accidental entry into a spin while at traffic pattern altitudes was not recoverable before ground impact. In addition minimal altitude loss was dependent upon both immediate pilot recognition and pilot recovery control technique.

The no-hands recovery was comparable to a normal recovery in terms of altitude losses. And while this technique has application at sufficient altitudes, recovery was not within the limits of traffic patterns altitudes. This technique, too, depends on pilot recognition of the spin and recovery technique. Since the no-hands technique works for spins up to three turns, the incipient phase for the Cessna 150, this technique depends on the pilot immediately recognizing the spin and direction of rotation. In addition, the pilot must also be proficient in the normal spin recovery procedures. While the no-hands technique may be used as an adjunct, it should not be substituted for the normal recovery when teaching spin recovery in a Cessna 150.

A total of 105 one-turn spins and 42 multiturn spins were performed for a total of 281 turns. In all spins both models of the Cessna 150 displayed no unusual flight characteristics and performed as indicated in the Cessna spin booklet. The Cessna A150M and Cessna 150L models were used because of slight differences in the tails. Although the tail configurations were identical, the vertical tail and rudder of the Model A150M was 6 inches longer than the Model 150L. One would expect the larger tail to be more effective in slowing the rotation and to thereby give a slight reduction in recovery turns. This was noted in the Cessna spin booklet, and this was found to be true for these two airplanes.

It was noted that the engine would stop in multiturn spins, but this was found when the fuel tanks were only one-fourth full. It was assumed that centrifugal force moved the fuel outboard in the tanks so as to unport the fuel intake. This assumption was supported by the fact that the fuel gauges indicated empty. The fuel sensor floats are located on the inboard side of the tanks and are parallel to the longitudinal axis of the airplane. Therefore, when fuel is displaced to the outboard side of the tanks, the fuel sensors indicate empty or near empty. When the tanks were one-half full there was a differential reading of the fuel gauges.

Within the limits of the investigation the Cessna 150 exhibited the spin characteristics established by the manufacturer. Recovery was predictable only when control input was in accordance with the procedure set forth by the manufacturer.

It should be noted that the investigation was limited to the incipient phase of the spin. Particular attention was directed to the one-turn spin to see if recovery was possible within the altitude limitations of the airport traffic pattern. The results indicated that recovery was not possible. This means that stall awareness and recognition is critical when flying in the airport traffic pattern.

The loss of altitudes for the incipient spin also indicates that stall practice should be done at a safe altitude above ground level. The three-turn spin begins the developed phase for the Cessna 150, and while the no-hands recovery technique worked during the incipient phase, the normal recovery technique was required for the developed phase.

Developed Spin

For research data on the developed spin in a Cessna 150, the *Basic Aerobatic Manual* by William K. Kershner was consulted. Kershner recorded the performance of a Cessna A150M (Aerobat) for a twenty-one turn spin. Rotation rate versus turns was illustrated in graphical form as shown in Fig. 9.2.

For this particular airplane the rotation rate increased until the fourth turn at which point the rate slowed. The rate began to increase at 7.5 turns. This cyclic pattern continued through the entire 21 turns, and at no time was the rotation rate constant. When the pitch attitude of the airplane was compared to the various turns and rotation rates, Kershner noted that the higher the pitch attitude (flatter spin) the slower the rates of rotation. Conversely, the lower pitch attitudes (steeper spin) had higher

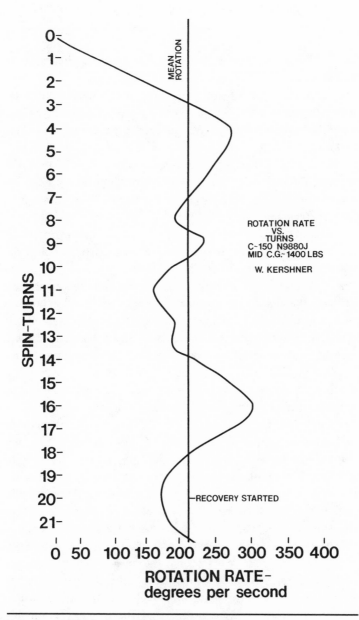

FIG. 9.2. Rotation rate versus turns for a
21-turn spin in a Cessna 150 Aerobat as
recorded by William K. Kershner. *The Basic
Aerobatic Manual* by William K. Kershner,
FIG. 3.4, p. 27. Reprinted with permission of
Iowa State University Press.

rates of rotation. During the first four turns the pitch attitude was steep as the rotation rate increased.

To more clearly illustrate the increase and decrease in rotation rate, straight line segments can be used to connect the major inflection points in the rotation rate versus turns curve. This is shown in Fig. 9.3. Line segments sloping to the right represent acceleration and those to the left deceleration. For example, the segment OA has an acceleration of approximately 76.5 degrees per second per turn. The segment BC shows a deceleration of 20.6 degrees per second per turn.

Two key facts need be noted about this research by Kershner. First, this was the reaction of a particular airplane loaded under a certain condition. Other airplanes, even of the same make and model, may have different spin characteristics. Second, the point in the cyclic pattern where a pilot initiates recovery may lead to deception for the pilot. For example, if recovery is initiated at point A, where the rotation rate is increasing, the recovery rate may be different from when recovery is initiated at point B, where the rate is decelerating. The pilot, sensing a difference in recovery rate, may not maintain recovery input long enough for recovery to take place.

In the research done by the author, the no-hands recovery technique did not work beyond the three-turn spin. Although this technique was not tested beyond the five-turn spin, Kershner's work offers a possible explanation. The mean rotation rate took place at the third turn and was not obtained again until 6.75 turns. It is possible that once the rotation rate reaches this value and beyond the elevator does not deflect enough, due to aerodynamic forces, to reduce the angle of attack and unstall the wings. In order to overcome the aerodynamic forces the pilot must apply force to the controls to keep the elevator deflected. Since the author's research dealt only in the area where the rotation rate was to the right of the mean rotation line (above mean value), it would be interesting to see what happened when the no-hands technique was used in the 6.75- to 8.25-turn range where the rotation rate was to the left of the

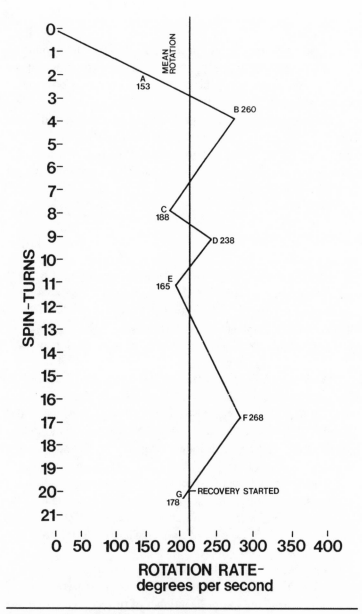

FIG. 9.3. Straight line segments connecting major inflection points in the rotation rate versus turns. Lines sloping to the right show increase in rotation rate, whereas lines sloping to the left represent a decrease.

mean rotation line (below mean value). To do this would be beyond the six-turn limit explored during the certification of the airplane and would place the pilot in the precarious position of being a test pilot.

If the above hypothesis is correct, the cyclic pattern of rotation rate versus turn could be used to predict at what number of turns a no-hands recovery would work, and where it would not, for a given make and model of airplane. This is certainly an area ripe for future exploration.

Manufacturer Testing

In 1975 the Cessna Aircraft Company conducted a flight test evaluation to determine if the spin characteristics of the Cessna 150 presented problem areas that might affect safety of operation. The investigation was conducted because two flight instructors had reported difficulties in recovering from spins.

TEST PARAMETERS. Four different Cessna 150 models were tested. The test primarily involved the Model 150K (17 spins) and Aerobat A150K (26 spins), but the Models 150L (14 spins) and 150M (5 spins) were also included. All were swept vertical-tail models, but the 150L had a 3-inch longer propeller shaft and a longer dorsal fin, whereas the 150M had a 6-inch longer vertical tail.

All test models were given ground conformity inspections to determine if control surface-rigging and weight-balance configurations were within manufacturer specifications. Also included were idle power RPM, nose gear strut extension, and range-of-control deflection.

Spin entries were initiated between 8,000 and 10,000 feet MSL. Variations in power and control positions were used for entry and during the spin. The number of turns ranged from one to six. Recovery procedures involved different inputs and sequences.

SPIN CHARACTERISTICS. The characteristics of the spin changed as the number of turns increased. Following entry the nose-down pitch was 50 degrees after 270 degrees of rotation. By two turns the nose-down pitch was 60 degrees and had increased to 70 degrees at three turns. At four turns the rotation rate had reached a maximum with a roll rate double the yaw rate. Spin rotation reached a steady state by five to six turns.

Addition of power with aileron into the spin increased rotation rate, particularly in roll. Addition of power with aileron against the spin flattened the spin (pitch of 45 to 60 degrees) and slowed the rotation. Addition of power with aileron neutral flattened the spin.

RECOVERY. Altitude loss was greatest for the one-turn spin (800 feet); however, the greatest loss per turn was in the three-turn spin. A four-turn spin required 1500 to 2000 feet for recovery, and a six-turn spin required approximately 2500 feet. The airspeed during recovery varied from 90 to 140 mph, accompanied by 2.5 to 3.5 G acceleration.

For recovery, rudder alone did not stop the spin or change the characteristics. Elevator alone gave a vertical pitch and increased rotation rate, and in some cases, a high-speed spiral resulted. Simultaneous use of rudder and elevator apparently blanked the rudder, delaying the recovery for up to three turns.

The manufacturer's recommended procedure of power idle, aileron neutral, rudder opposite rotation followed by forward elevator gave the most consistent recovery. Neutral aileron was critical since the ailerons were aerodynamically effective during the spin. Aileron input altered rotation rate and pitch so as to delay recovery.

PILOT EXPERIENCE. The two flight instructors reporting the difficulties in spin recovery were interviewed. It was determined that the spins had been entered from a low altitude (3000 to 4000 feet AGL) and were terminated at two to three turns. The instructors did not know the effect of aileron input on the spin and did not know the importance of control input sequence for recovery. Proficiency in prolonged spins with high rotational rates was questionable.

SUMMARY. Based on the flight test data, Cessna Aircraft Company made the following recommendations:

1. Airplanes used for spin maneuvers should be checked frequently for correct tolerances on control system rigging. Nose gear strut extension should be checked to assure proper locking of the nose gear when in flight.

2. Accurate weight and center of gravity positions should be checked for airplanes used in spin training, especially when additional equipment has been installed. Any mass distributed at the extreme ends (100 inches or more aft of the datum) of fuselage produces a moment of inertia that causes the pitch to decrease whenever the airplane is spinning in a steep pitch down attitude. Fuel distribution at spin entry will alter spin characteristics.

3. Pilots should be aware that spin rate increases and pitch approaches vertical as the number of turns increase. Rotation rate is a direct function of pitch-down attitude: steep pitch attitude, high rotation rate; shallow pitch attitude, slower rotation rate. This phenomenon tends to disorient pilots.

4. Manufacturer recommended recovery procedure gave the most consistent and effective recovery.

Summary

Spin characteristics of the Cessna 150 have been discussed at some length for three reasons: First, the Cessna 150 has and continues to be the most widely used primary training aircraft; second, the Cessna 150 is certified for six-turn, intentional spins; and third, there is documented test data available on spinning the Cessna 150.

Test results show the Cessna 150 to have predictable spin characteristics provided the airplane is maintained and flown within manufacturer recommendations. Despite this predictable spin mode, a pilot must operate the airplane within the realm of personal expertise based on a solid foundation of spin aerodynamics and human factors relative to a spinning airplane.

10
Human Factors in the Spin

Human Factors

The factors associated with the aerodynamics of a spinning airplane have been and continue to be an area of concern; however, in this effort to determine the many factors associated with the aerodynamic spin, one major component has been largely ignored, the human factor. The human problems associated with the spin are as critical to spin recovery as the design of the airplane. The airplane can be designed to have certain predictable spin characteristics, but design alteration is not possible for the pilot; thus the pilot must be aware of the human factors involved in an aerodynamic spin. While a pilot may be proficient in spin recovery techniques, these techniques may be of no use if the pilot is disorientated with respect to his environment.

Sensory Mechanisms

To appreciate this human problem of disorientation it is necessary to first understand how the pilot obtains information about his environment. A pilot's primary internal sources of information are three sensory mechanisms: the proprioceptive, the visual, and the vestibular. The information from these three sensory mechanisms is integrated and formulated through perceptual processes so that the human organism functions with a continuous picture of the environment. Consequently, during the process of human maturation, human perceptions are based on a gravity-constant environment. In addition, certain sensory information leads to the development of reflex patterns of response that are relative to a constant gravity. For example, when a person stumbles, neuromuscular responses of reflex origin come into play to maintain body equilibrium with respect to the earth's gravity.

On earth, human rotational movement is relatively slow and usually limited to motion about only one axis of the body. For such movement the visual and vestibular sensations are reliable; however, when the human is placed in the three-dimensional realm of flight, earth-related visual references may not be available. Gravitational pull may not be aligned along the usual body axis, and the force of gravity may be altered by the acceleration of airplane maneuvers.

Proprioceptive sensors provide the pilot with information about joint position and muscle tension, but, compared to visual and vestibular information, proprioceptive sensation is a small part of the total experience. The vestibular apparatus is the sensory organism that detects change in motion relative to an equilibrium position. From a comparative viewpoint when in flight, visual sensation is the most reliable, proprioceptive is less reliable, and vestibular is very unreliable, yet very sensitive and powerful. Consequently, when in flight, spatial disorientation is usually produced by a conflict between visual and vestibular sensation.

Vestibular Apparatus

The vestibular apparatus, located in the middle ear, detects change in motion through two structures—the utricle and the semicircular canals. The utricle detects linear acceleration and

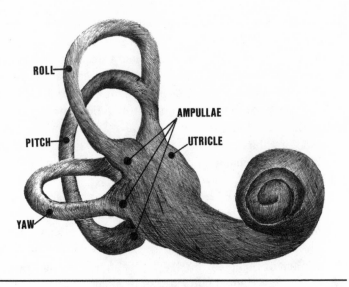

FIG. 10.1. The vestibular apparatus detects change in motion through two structures, the utricle and the semicircular canals.

the semicircular canals determine angular acceleration (Fig. 10.1).

UTRICLE. The utricle is a small saclike structure located in close proximity to the semicircular canals. When the head is tilted this sensory mechanism indicates the direction and magnitude of the linear acceleration being experienced by the head. This allows the position of the head relative to the earth to be constantly monitored. The pilot is able to detect level flight (both upright and inverted), banked flight, and climb and descent.

SEMICIRCULAR CANALS. The three semicircular canals are structured with the canals perpendicular to each other so that a canal lies in each of the three principal planes of the human body (Fig. 10.2). The canals are filled with a fluid (endolymph), and the motion of this fluid is detected by sensory hair cells located in an

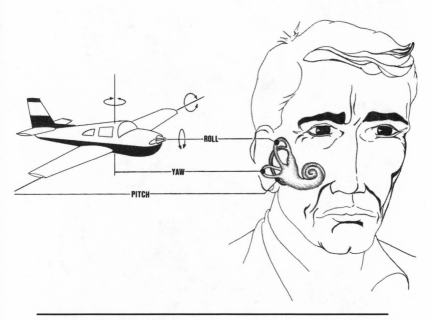

FIG. 10.2. The semicircular canals are structured with the canals perpendicular to each other so that a canal lies in each of the three planes of the human body.

enlarged area (ampulla) at the end of each canal (Fig. 10.3). When the head begins to rotate in any direction, the fluid inside the appropriate canals tends to remain stationary because of inertia. This results in a relative motion of the fluid opposite to the direction of head rotation. Once the head is rotating at a steady velocity, the fluid, because of friction, now moves at the same velocity as the canals. Fluid movement displaces the hair cells which in turn send impulses to the brain. Based on this information the brain determines the direction of rotation. Since each canal lies in a different plane, the semicircular canals can report rotation in three dimensions.

Once the rate of rotation becomes constant the fluid and canal rotate at the same rate. The hair cells are no longer stimu-

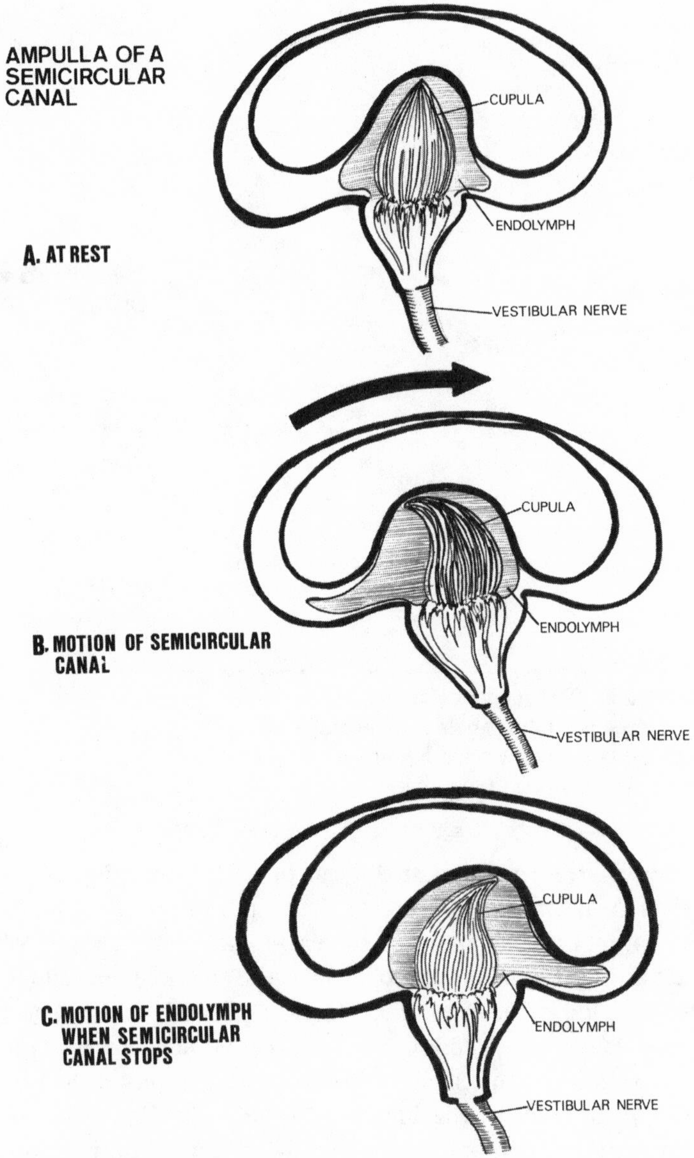

FIG. 10.3. (A) With the head at rest, the
endolymph is stationary. (B) When the head
rotates to the left, there is relative motion of
the endolymph to the right. (C) The
endolymph rotates in the opposite direction
when rotation of the head stops.

lated and the brain interprets this as no rotation. However, when the canals actually stop rotating, the fluid continues to rotate and the hair cells are bent opposite to the direction of original rotation. The brain interprets this as rotation opposite to the direction of original rotation.

For the pilot, this creates a confusing situation called the *coriolis illusion* or cross-coupling (Fig. 10.4). If the airplane is in a constant rate turn and the pilot moves the head in any plane of motion different from that of the airplane turn, he will believe that the airplane is turning in a different direction from what is actually taking place. For example, if an airplane has motion about the pitch and yaw axes and the pilot rotates his head

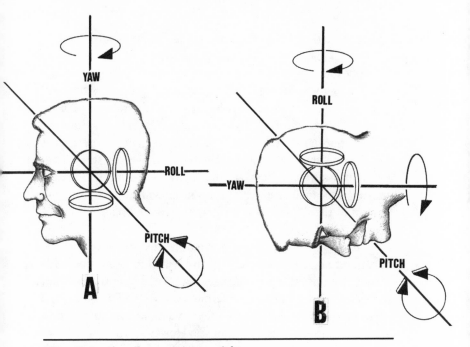

FIG. 10.4. If a pilot experiences right yaw (A) and then tilts his head ninety degrees in pitch, he will experience motion in roll (B) due to cross-coupling.

HUMAN FACTORS IN THE SPIN

forward 90 degrees about the pitch axis, the pilot will experience a sensation of motion about the roll axis.

SUMMARY. In summary, the semicircular canals detect angular acceleration, the rate of change of rotation of the head in any direction. The angular acceleration required to stimulate the semicircular canals is about 1 degree per second per second. The canals do not detect static position or linear acceleration. Depending on the circumstances, information from the semicircular canals can be misleading.

Visual Tracking

Information from the semicircular canals is important to visual tracking. When the head rotates the eyes must rotate in the opposite direction if they are to remain fixed on an object long enough to produce a clear image. When the eyes have rotated to the limit, they move rapidly in the direction of head rotation in order to fix on a new object. They then again move slowly opposite head rotation. This movement of the eyes is called *nystagmus*. When the semicircular canals are stimulated, nystagmus always occurs automatically. Any time the head is subjected to angular acceleration, an immediate compensatory motion of the eyes takes place in the direction opposite to rotation of the head.

When functioning in a gravity-oriented environment the human organism integrates the information from the proprioceptive, visual, and vestibular sensory mechanisms so that the eyes are able to produce intermittent stabilized images of the environment in order that the neuromuscular system can provide the appropriate body responses. However, when the human organism is placed in flight environment, the semicircular canals frequently misrepresent the rotational motion of flight. The eyes, through visual interpretation of the environment either through outside references or instruments, become important as means of orientation in flight.

The aerodynamic spin is a flight condition that has the potential of producing incorrect representation by the semicircular canals, a condition that the eye cannot correct. The spin includes rotational motion of roll and yaw in the planes and usually includes a pitching motion. This motion is frequently oscillatory in at least one plane. Consequently, the pilot is subjected to rotational stimuli in all three planes of the semicircular canals.

To determine the difficulties the aerodynamic spin imposes on the vestibular and visual perceptual mechanisms, G. M. Jones of the British Royal Air Force Institute of Aviation Medicine conducted an investigation of the compensatory eye movements in the three planes of yaw, pitch, and roll when the pilot was subjected to the angular motion of a spinning airplane. The findings have significant implications for a pilot who engages in the spinning of airplanes.

In the yaw plane the airplane angular velocity rose to a peak of 50 to 60 degrees per second. Correct compensatory eye velocity was maintained until 20 seconds, then became intermittent. On recovery, slight reversed nystagmus was evident. The airplane pitch motion was oscillatory, the motion being such that the average angular velocity was zero. This allowed both the vestibular and visual mechanisms to correctly orient the pilot. On entering the roll plane, the airplane accelerated rapidly to about 150 degrees per second. At this point the motion became oscillatory, but the mean angular velocity continued to increase. The eye was able to compensate for the motion during the initial portion of the roll, but as the motion continued the eye motion began to lag until at recovery the eye motion was reversed. Consequently, the pilot was unsure of his orientation and unable to determine spin direction.

Interestingly, the eye motion was the same regardless of whether the pilot looked outside the cockpit or inside the cockpit at the instruments. Since the instrument panel was moving the same as the pilot's head, one would expect the eye motion relative to the pilot's head to be zero. This was not the case. Looking at the instruments did not diminish the disorientation for the pilot.

In summary, during the initial phases of the spin the vestibular-induced compensatory eye movement is appropriate for all three planes. The eye is capable of maintaining an intermittent retinal image of sufficient stability so that the pilot is oriented to his environment. However, after about five turns the eye movement is out of synchrony with the airplane rotation. The pilot now has a blurred, streaked visual reference, and the spin appears to have increased in speed. It is theorized that the lack of visual stabilization is due to decreased signals from the semicircular canals. At this point the pilot has difficulty determining the number of turns in the spin.

The discrepancy between the visual and vestibular mechanisms is dependent on the nature of the oscillatory motion. If the oscillating motion produces an average velocity of zero there is minimal problem with visual stabilization. However, if the average velocity is not zero and continues for a sufficient duration, reversed nystagmus takes place. In the spin, the roll plane appears to be more affected than the other planes.

To some extent it is theoretically possible to offset the physiological problems found in the roll plane. By directing visual scan on the actual horizon during the spin and recovery, the rotational motion is placed in the yaw plane rather than the roll. However, it is difficult and at times impossible to locate the horizon. In addition, any motion of the head relative to the airplane produces cross coupling and contributes to disorientation.

It is in the critical recovery phase where events become confusing for the pilot (Fig. 10.5). Suppose the airplane has been spinning to the left. Control input from the pilot stops the rotation, but in the semicircular canals the fluid, because of inertia, continues to move to the left relative to the canal. This is

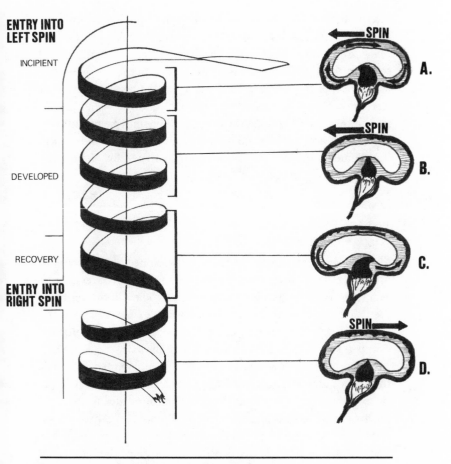

FIG. 10.5. Entry has been made into a left spin. (A) The endolymph lags behind the motion of the semicircular canal. (B) The motions of the endolymph and semicircular canal are in equilibrium. (C) Spin rotation has stopped but the endolymph continues motion in the direction of the spin. (D) Because of conflicting sensory information, the pilot enters a second spin in a direction opposite the first.

interpreted by the pilot's brain as rotation to the right. However, with the compensatory eye motion reversed such as in the roll plane, the visual reference indicates a rotation of the visual scene to the right relative to the pilot. The brain interprets this as rotation to the left. Thus there is a conflict between the semicircular canal indication of turning right and the visual indication of turning left despite the fact that there may be no rotation at all. This accounts for how a pilot can enter a second spin at the moment of recovery from the first. Generally, the second spin is opposite in rotation to that of the first.

G-Load

The spin, from entry through the developed phase, does not usually result in high acceleration (G). During recovery, however, the G-load may become quite high. The pilot determines the G-load when he pulls out of the dive following spin recovery. The airplane's resistance to G-load-induced structural deformation is a function of the category design limits.

The human, however, has a variable but limited tolerance to G-load, and, if the tolerance is exceeded, the pilot will experience loss of consciousness (LOC). The pilot, during the pullout from a spin recovery, experiences G-load along the head-to-foot axis, and, because of the vertical orientation of the blood vessels, there is increased hydrostatic pressure of the blood that causes the blood to pool in the lower torso and legs. The resulting deprivation of blood to the brain causes LOC.

Many physiological factors affect a pilot's susceptibility to G-load-induced LOC. Body size, physical fitness, activity level, fatigue, and medications are but a few such factors. Besides physiological factors, the nature of the G-load is also involved. The effect of G-load on the pilot depends on such parameters as magnitude, duration, and rate-of-onset (Fig. 10.6). For example, a high magnitude G-load of long duration may cause LOC,

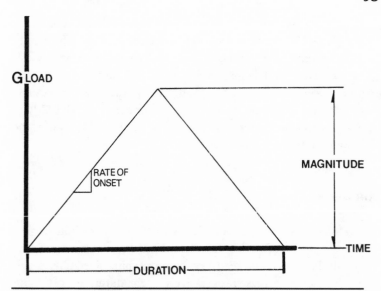

FIG. 10.6. The effect of G-load on the pilot depends on magnitude, duration, and rate of onset of the load.

but the same magnitude for a short duration will not give the same results.

Because of the relatively low magnitude of the G-load during spin recovery, most pilots do not consider the possibility of LOC; however, a study by J. E. Whinney, U.S. Air Force School of Aerospace Medicine, shows that LOC is a significant factor during spin recovery. The purpose of the report by Dr. Whinney was to document the flight operations that produced G-load-induced LOC during military undergraduate pilot training. During a 42-month period there were 69 reported LOC episodes, and the spin/dive recovery accounted for 16 (23 percent) of the total. This ranked second in the 11 different aerobatic maneuvers. The mean G-load was +3.8, and the mean time of total incapacitation was 12 seconds.

The 3.8-G average reported in this investigation is within

HUMAN FACTORS IN THE SPIN

the design load limits of general aviation airplanes. Such a G-load is highly probable when spin recovery is in close proximity to the ground. Furthermore, with the combination of physiology variables and G-load parameters, it is difficult to predict pilot susceptibility to LOC. During any spin recovery, regardless of whether the pilot be experienced or a novice, the G-load limitation of the pilot is just as essential to safe recovery as is the design limit of the airplane.

Summary

In view of the physiological structure of the pilot, human factors can be of major consequence in the aerodynamic spin and in the subsequent recovery from the spin. Since physiological difficulties increase progressively as the duration of the spin increases, it is important that spin maneuvers be confined to approximately 5.0 seconds, particularly for the novice pilot. During this time frame the pilot can direct his attention to the aerodynamics of the spin rather than the physiological problems. With repeated practice the pilot is better able to cope with the physiological problems of the spin, but this habituation is rapidly lost. Therefore, even for the experienced pilot, it is advisable that spins be approached in stages so as to permit habituation. Certainly a flight instructor must understand the human factors in the spin when teaching spin entry and recovery.

11
Spin Accidents

Accident Investigation

While accident investigation is not the solution to the problem of airplane accidents, it is the keystone for air safety. The basic purpose of an investigation is determining cause in order to prevent similar accidents. This premise is not new, but seldom are investigative results exploited to their full preventive potential.

There are two limitations in the accident investigation-prevention relationship. First, the investigative results must be available to the pilot, and the pilot must incorporate the preventive recommendations in his flight operation procedures. Second, the predominant factor in accident causation, the human factor, is the most difficult to investigate. It is not enough to determine that the pilot made an error. Before any corrective or preventive action can take place, it must be learned why the pilot made an error. While the precise reason can seldom be determined, hypothetical explanation can still be of safety value.

The investigation of a spin accident is difficult, primarily because of the human element involved. Even with no witnesses, it is not difficult to determine that an airplane spun in and crashed. The wreckage distribution usually follows a characteristic pattern (Fig. 11.1). However, the human activity that took place from the time the airplane entered the spin mode until the descent abruptly ended with ground impact is a matter of speculation because a fatality is usually the end result. Hence, hypothetical explanation is the only alternative.

The following spin accidents were taken from the NTSB files. The three distinct types of accidents were selected to illustrate how human factors combined with particular design characteristics of an airplane can, under certain conditions, result in a spin accident.

Instructional Spin Accident

ACCIDENT INVESTIGATION. A flight instructor and student departed the airport in a Piper Tomahawk (PA-38) for a 1-hour VFR instructional flight. The fuel tanks were topped off, a weather briefing was obtained, and a VFR flight plan was filed.

FIG. 11.1. Reconstruction of a light-airplane spin accident. Aircraft Accident Investigation Laboratory, Institute of Safety and Systems Management, University of Southern California.

The airplane was observed by witnesses in a designated training area in level flight with engine noise being steady. The engine noise decreased just before the nose dropped and the airplane began a spin to the left. An estimate of altitude could not be made.

The airplane impacted the ground in a nose-low, left wing–low attitude. The surrounding tree and airplane damage indicated a near vertical descent with left rotation. The vertical speed indicator showed a 2,000 fpm descent and the needle was pegged. The tachometer read 850 rpm and the airspeed indicator showed 94 knots. The carburetor heat was "on" and the fuel boost pump was "on." Both fuel tanks were damaged and empty, but rescue persons indicated that a strong fuel odor was present. Examination of the engine and power controls revealed no problems other than impact damage. The flight-control system showed no evidence of malfunction.

The weight and balance was calculated both for takeoff and at the accident site. The weight was 58 pounds over the maximum gross weight (1,670 pounds) at the time of the accident. The center of gravity (CG) was within the operational envelope.

Both pilots were fatally injured by the impact. Postmortem examination showed no medical reason for incapacitation of either pilot prior to the impact. Comparable fractures of the hands, particularly the thumbs, were found on both pilots. Toxicology tests done by the Civil Aeromedical Institute (CAMI) were negative for drugs, alcohol, and carbon monoxide, except the instructor's urine contained phenylpropanaline. The instructor was noted to be obese.

The instructor was employed full time by a flying club and was considered to be an excellent teacher. He was efficient and methodical in his instructing and used a maneuver list for each lesson. Such a list was found in the airplane wreckage. The instructor had received formal spin training in a Tomahawk. Although he received above average grades, the content of the training was not given.

For the three nights preceding the accident the instructor

had averaged 6.5 hours of sleep rather than his usual 8 hours. He had logged 4.9 hours of instruction two days prior and 8.4 hours the day before the accident. The accident flight was the third of that day. He was in a jovial mood prior to the accident flight.

The student pilot had recently completed military helicopter pilot training, and held the civilian commercial pilot certificate for helicopter with instrument rating. The student had had approximately 4 hours of fixed-wing instruction at the time of the accident. The student had received stall and spin instruction on his second flight.

The NTSB listed two factors as probable cause: (1) the student failed to obtain and maintain flying speed; and (2) the instructor inadequately supervised the flight maneuver.

ANALYSIS. Accounts by the witnesses, instrument readings, and the wreckage pattern confirmed a stall/spin accident, but only a hypothetical explanation can be determined.

The Piper Tomahawk (Fig. 11.2) is a two-place trainer designed as a teaching tool and is certified for spins. It has a T-tail, low-wing with rectangular (Hershey Bar) planform, and GAW-1 airfoil. The airplane does have a high percentage of stall/spin accidents.

There is ample buffet to warn of an approaching stall, but the stall itself is abrupt with no two airplanes having the same stall pattern. At stall there is quick roll-off in one direction or the other. Unless prompt and positive rudder is applied the airplane easily and quickly enters a spin. This loss of lateral control at stall is surprising based on the wing planform. With a rectangle wing the stall progresses from the wing root to the tip. The lift generated at the ends of the wing versus the lift loss at the roots provides for lateral stability at the stall break.

The slender GAW-1 airfoil produces a high lift coefficient, but small changes in the curvature of the airfoil significantly alter the lift distribution. The Tomahawk wing is built with five full-chord-thickness ribs. The skin covering the wings has metal

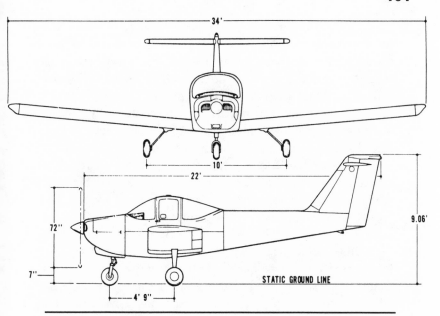

**FIG. 11.2. The Piper Tomahawk (PA–38) is
a two-place, T-tail, low-wing trainer.
Courtesy of Piper.**

strips located between the ribs to "stiffen" the skin. However, the
skin has considerable flex or "oil canning" that changes the curvature of the wing. This "oil canning" is evident during flight.
As the wing approaches the critical angle of attack the pressure
differential caused by the reduced airflow over the wing produces an asymmetrical curvature which gives an unequal lift
pattern over the two wings. The resulting asymmetrical lift produces an inconsistent roll-off at stall.

Not only is the roll direction unpredictable, but the degree
of roll varies from 30 degrees to as much as 90 degrees. The
addition of stall strips (two sets) was intended to counter the
abrupt stall conditions and roll characteristics, but the effectiveness has been questionable. The roll at stall is still unpredictable.

SPIN ACCIDENTS

With the abrupt stall and unpredictable high degree roll, the Tomahawk will quickly enter an inadvertent spin. The rotation rate increases rapidly with a steep, nose-down pitch attitude. Airspeed stabilizes at approximately 85 knots, well above stall value. Full rudder application opposite rotation followed by full forward yoke is required for recovery. Recovery requires up to two turns and a one-turn-spin altitude loss is 1000 to 1500 feet. The fast, steep spin mode can easily disorient the pilot.

With the T-tail configuration and large rudder, one would expect the Tomahawk to have a quick recovery from a spin since the rudder is not shielded from the relative wind by the horizontal tail. However, it seems that the rudder is shielded from the relative wind by the wings when in the steep spin mode.

Although the weight and balance were within the CG envelope for the Tomahawk involved in the accident, the overweight condition altered the mass distribution. This change in moment of inertia would delay the recovery from a spin, especially as the number of turns increased and a developed spin mode became established. In this accident the airplane was overweight.

As the control yoke is moved to the full aft position, the direction of travel is up as well as aft. When in the aft position any force applied in the vertical direction prevents the yoke from moving forward. Hence a pilot pushing forward and up on the yoke would believe the control to be jammed.

The instructor's maneuver list, found in the wreckage, contained three maneuvers that could readily lead to a spin; minimum controllable airspeed, approach-to-landing stall, and takeoff–departure stall. The carburetor heat "on" suggests a forced landing procedure, but the control lever is located in a position that is not visible to the instructor; therefore, it could have been left on from a previous maneuver. Carburetor heat and boost pump are routinely used in power-off maneuvers. The decrease in engine noise (power reduction) followed by the spin entry suggests an approach-to-landing stall maneuver.

Despite the student's helicopter experience, he was a beginner with respect to stalls and spins. Although the instructor had spin training for the instructor certificate (and specifically in the Tomahawk), he was not necessarily experienced in multiturn spins.

FAR 61 does not specifically state that an instructor applicant must demonstrate spins on the practical flight test. The instructor applicant must be able to recognize an imminent spin entry and promptly recover the airplane to normal flight. The applicant may be asked to demonstrate recognition and recovery from spin situations that may be encountered in poorly performed maneuvers during student-pilot flight training. However, the examiner conducting the test may accept a logbook entry of spin competency rather than require an actual demonstration. FAR 61.45 requires the logbook entry to be certified by the flight instructor who conducted the spin training. Therefore, the flight instructor applicant may have never been beyond a one-turn spin and this may have been endorsed by his instructor—who had never been beyond a one-turn spin.

It is possible for an instructor to have considerable experience in the mechanism of spin entry and one-turn recovery and yet have only a basic understanding of what is actually happening to the airplane and the pilot during the entry and recovery. In the multiturn spin, instructor expertise is usually less than marginal at best. Very few pilots understand the relationship between airplane design factors and spin mode. This could well have been the case in this particular accident.

The instructor was not in top physical condition. His normal sleep pattern had been disturbed for three consecutive nights prior to the accident. In addition, his workload had been heavy, particularly the day before the accident. There is no doubt that the instructor was fatigued. Finally, the phenylpropanaline found in his urine could have contributed to the accident.

Phenylpropanaline is a common ingredient in antihistamine/decongestive medication sold over the counter. The drug is not an antihistamine but a nasal decongestant and appetite suppressant. It is a central nervous system stimulate that can easily be overdosed despite being 80 to 90 percent excreted from the body within 24 hours. Overdose side effects may cause dizziness, difficulty sleeping, and an unusual feeling of well-being.

No evidence of allergy or respiratory problems were noted, but the postmortem report did note that the pilot was obese. Hence the drug may have been for purpose of appetite suppression. Regardless, the drug could have contributed to poor rest during a disturbed sleep pattern. The drug could also have contributed to disorientation during the spin mode.

The fracture pattern in the hands, especially in the thumbs, of both pilots indicates both had their hands on the controls at the time of ground impact. Injuries to the lower extremities could not be associated with rudder/foot relationship because of the severe deformation of the fuselage at ground impact.

CONCLUSION. Due to the human factors involved in the accident, the conclusion must be hypothetical. This, however, does not negate the accident prevention potential of the conclusion.

The Tomahawk entered the spin from an approach-to-landing stall maneuver. The abrupt stall accompanied by the unpredictable roll characteristic of the Tomahawk, coupled with the inexperience of the student pilot, led to an inadvertent spin entry to the left.

The instructor, not at his physical best due to fatigue and the effects of medication, did not promptly react to the spin entry. This delay was critical considering the quickness with which a Tomahawk enters the spin mode. The delay was also critical because of the altitude loss associated with the spin. Since the impact airspeed was above the spin mode airspeed, recovery was apparently initiated.

Since both pilots had their hands on the control yoke, either there was disagreement as to the recovery input needed or the control appeared to be jammed. The first could be due to disorientation on the part of the inexperienced student pilot and to the combined fatigue/drug effects on the instructor. The second could be due to a vertical force rather than to forward pressure applied to the yoke. Mass distribution due to the airplane being overweight also delayed the recovery. All of the above, combined with the spin characteristics of the Tomahawk, resulted in a recovery delayed too long to avoid impact.

Go-Around Spin Accident

ACCIDENT INVESTIGATION. A Cessna 150 with two private pilots on board was landing at a small airport (runway length 3000 feet, width 60 feet). There was a 5 knot, 10 degree, left crosswind. The final approach was high and when the airplane drifted to the right of the runway upon flaring, the pilot initiated a go-around. The go-around was witnessed by a pilot taxiing to the runway.

Just after the airplane flared to level, the nose pitched up to a steep climb. The fully extended flaps remained down as the airplane climbed. At an altitude of approximately 100 feet AGL the airplane stalled and turned to the left. After turning approximately 100 degrees to the left the airplane impacted the ground in a nearly vertical descent. The left wing impacted before the right wing made ground contact. The flaps were in a 40-degree down position.

Postmortem examination on the pilots was negative for drugs and alcohol. No evidence of incapacitation prior to ground contact was noted. Injuries were too severe to determine who was flying the airplane at impact.

Both pilots were experienced in the make and model. Although logbooks were unavailable it was reported that the pilots had 700 and 450 hours of flight time with biannual flight reviews (BFR) in the Cessna 150 completed 6 and 13 months prior to the accident. The pilots were co-owners of the airplane and frequently flew from the airport. Both were considered of average ability but cautious. It was noted that neither pilot routinely used 40-degree flaps for landing.

Engine and airframe investigation revealed nothing to indicate either engine or control-system malfunction. Although the fuel tanks ruptured, there was evidence of fuel at the accident site. The instruments were not readable but the throttle was full open and the carburetor heat was off. The trim wheel was aft (nose-up) of neutral. The airplane was within CG and weight limits at the time of the accident.

The NTSB found the causes of the accident to be failure to maintain airspeed and to correctly use the airplane equipment.

ANALYSIS. The pilot failed to properly execute a go-around, allowing the airplane to enter a left spin from a take-off–departure stall. The reasons for the failure are unknown, but an analysis of the accident has safety value.

The Cessna 150 is a two-place, high-wing, conventional-tail airplane with a fairly complex flap system for a flight trainer. The slotted Fowler flap can be extended to any degree of deflection from 1 to 40 degrees (Fig. 11.3). As the flaps deflect down they also move aft 8 inches when fully extended. Early model Cessnas have a manual flap lever whereas the later models use an electrical motor that requires 9 seconds for full extension and 6 seconds for retraction. The flaps are large, representing 11.5 percent of the total wing area, 17.1 percent of the wing span, and 30 percent of the wing chord.

Full flap extension decreases the stall speed 12.7 percent. The first 20 degrees of extension lowers the stall speed 6 mph while the second 20 degrees lowers it only 1 mph. The first 20 degrees of deflection primarily produces an increase in lift, but the next 20 degrees produces a high drag condition. The airflow over the top of the flap begins to separate and the drag increases. The slotted Fowler delays this separation, decreases the critical angle of attack approximately 2 degrees, and lowers the stall speed. The flaps are parasite drag, and this drag, is proportional to the square of the airplane speed. This drag condition allows a steep approach that is excellent for landings, particularly over an obstacle, but hinders take off and go-around.

With 40 degrees of flaps, a go-around requires immediate

FIG. 11.3. The Cessna 150 has a slotted
Fowler flap that can be deflected to 40
degrees.

20 degrees of retraction following application of full power. Retraction of 20 degrees reduces drag and allows the airplane to accelerate. Airplane pitch is held in level flight to allow the airspeed to build. As the airspeed increases the flaps are retracted in increments. This prevents a loss in lift that would allow the airplane to settle, possibly allowing ground contact. Complete retraction of 40 degrees of flaps, despite requiring 6 seconds, would lead to a rapid loss of lift and, depending on the angle of attack, result in a stall.

The high-wing location of the flaps causes the airplane to pitch up as the flaps are extended. The downwash from the deflected flaps increases the airflow over the tail, creating a nose-up moment. The pitch increase is very prominent with the application of full power since the prop wash (airflow) increases along the fuselage. Forward pressure is needed on the yoke to keep the nose from quickly pitching up to the critical angle of

108

attack. Right rudder is needed to offset the left turn induced by engine torque, P-factor, gyroscopic action of the prop, and prop slipstream.

The Cessna manual indicates that normal landing approaches can be made with flaps up or down. When landing in a strong crosswind, the minimum flap setting required for the field length is to be used. For short field landing an approach with 40 degrees of flaps is recommended, and touch-down with power off is recommended.

CONCLUSION. The accident scenario is fairly evident in this particular case. A take-off–departure stall with a left spin entry resulted from a poorly executed go-around.

The pilot was making a short field-landing approach to the airport with a slight left crosswind. The final approach was made with 40 degrees of flaps, a configuration not normally used by the pilot. The approach was at 60 to 70 mph with the trim wheel set aft (nose-up) of neutral. Because of the crosswind the airplane drifted right-of-center-line, and the pilot initiated a go-around.

The application of full power produced a pitch up attitude due to the 40-degree flap setting and the aft trim. The full power setting caused the airplane to turn to the left. Neither of these situations were countered with the correct control input. In addition, the flaps were left fully extended, causing the airspeed to decrease rapidly as the pitch increased. The stall with left yaw produced a spin entry to the left. Altitude was not sufficient for recovery.

Pattern Spin Accident

ACCIDENT INVESTIGATION. A commercial pilot was practicing touch-and-go landings in a Piper J-3 Cub (PA-11) at a small airport with a turf runway. There was a 20-degree, 12-knot, left crosswind. When turning from base to final the airplane pitched down and rolled to the left. After approximately 180 degrees of rotation the airplane impacted the ground, fatality injuring the pilot.

A pilot waiting to take off witnessed the turn from base to final and observed that the airplane was "moving sideways" just prior to pitching down and rolling to the left.

Examination of the wreckage failed to find any malfunction with the engine or the airframe. Postmortem analysis on the pilot was negative for drugs and alcohol, and there was no evidence of incapacitation prior to impact. The pilot had logged over 2000 hours, most in single engine Cessna airplanes, and had nearly 50 hours in the J-3 Cub.

The NTSB listed the cause of the accident as failure to maintain airspeed.

ANALYSIS. The J-3 Cub can be considered a simple two place airplane, and it is simple as far as construction and systems (Fig. 11.4). Original instrumentation was airspeed, tachometer, alti-

FIG. 11.4.   The Piper J-3 Cub (PA–11) is a simple two-place airplane. Note the geometric design of the rudder.

meter, slip (ball) indicator, compass, oil pressure, and temperature. With no vacuum system there is no artificial horizon and directional gyroscope, and with no electrical system there is no turn coordinator. Hence the pilot must fly strictly by pitch attitude and kinesthetic sensation.

The tandem-seat Cub (solo from rear seat) has long wings (35 feet 2.5 inches) and wide wings (chord 5 feet 3 inches). There are no differential ailerons so that aileron input for roll also results in the nose pointing in the opposite direction unless the adverse yaw is offset with the rudder. The input required is more pronounced than in modern-day airplanes.

The fairly large rudder (6.55 square feet) is capable of producing large sideslips. The upper portion of the rudder overhangs the vertical stabilizer. In other words, it is on the opposite side of the rudder hinge line. This acts like a "spade" found on modern-day aerobatic airplanes. As the control surface is deflected the rudder overhang assists movement of the rudder because of the aerodynamic force on the overhang. Hence, as the airspeed increases, the rudder control pressure gets softer and movement becomes easier.

The Cub is very maneuverable, with low speeds and tight turning capability. The stall speed is 38 mph at 0 degree bank, but increases to 54 mph at a 60-degree bank. Normal approach to landing is made at 50 to 60 mph, depending on load and wind gust conditions, with most approaches made at 55 mph. It is extremely light (approximately 800 pounds) and is very sensitive to turbulence and wind.

The Cub will spin easily and recover readily but demands altitude. While documented data is lacking, 800 to 1000 feet are needed for recovery. The Cub has a high rate of stall/spin accidents.

Because of the airplane's simplicity, pilots of modern-day airplanes tend to equate the simplicity with ease in flying. Lack of understanding of the control system frequently results in poorly coordinated turns. Without instruments for reference, the airplane banks to a greater degree than realized. The probability of making this mistake is increased since modern pilots are accustom to side-by-side seating rather than tandem. The solo from the rear seat also places the pilot further back from the cowl and instrument panel than when flying a modern airplane. In essence, the pilot has to rely on pitch attitude solely from visual perspective, something for which pilots are not trained in most modern-day airplanes.

CONCLUSION. There is no question that the Cub was in a cross-control turn that developed into a spin under the bottom.

Because of the left crosswind the pilot overshot the turn from base to final. In an effort to align the airplane with the runway, left-rudder input was increased but the right aileron was used to prevent overbanking. Back pressure on the stick was used to keep the nose from dropping. Accustomed to side-by-side seating and cross-reference of instruments with pitch attitude, the pilot misjudged the configuration of the airplane. Contributing to the misjudgment was the pilot's lack of familiarity with the pitch-attitude view from the cockpit of the J-3 Cub as compared to the Cessna airplanes that he primarily flew. The critical angle of attack was exceeded (stall) and when combined with the left yaw (rudder deflection) the result was a spin under-the-bottom.

Summary

There were common elements in each of the above accidents. The airplanes were two-place and considered to be trainers, a status usually associated with simplicity of flight operation. Although the pilots had different levels of experience, each was involved in a stall/spin accident. In each case the NTSB listed the failure to maintain airspeed as a cause. In actuality there was not a single cause, but multiple causes that took place in a particular sequence, the end result being a stall/spin that resulted in an accident.

The specifics of the accidents have been presented. Although only hypothetical explanations were set forth, there are general lessons that can be learned. First, the pilot must know more aerodynamics than just the fundamentals of lift and drag. The spin is not a normal flight mode, but the aerodynamics of entry and recovery are relatively simple. Second, the pilot must understand the relationship between basic principles of airplane design and flight characteristics. In this way a pilot knows what to expect in certain flight operations. An aeronautical engineering degree is not required to understand the basic relationships. Third, the pilot must be aware of the different accident scenarios that can take place during a particular flight operation. A standard procedure for each phase of flight operation, strictly adhered to, will prevent a sequence of events from developing that will lead to an accident. Fourth, the pilot must know himself or herself. The pilot is an integral part of the airplane. Just as the airplane has limitations, the pilot also has limitations. While some limitations result from deficiencies in training and ability, others are physiological and psychological traits of all humans. Unfortunately human limitations are not as well defined as mechanical limitations. It is the pilot's lack of knowledge about human limitations that is the primary cause of pilot error. Accident investigation can provide causes, both factual and hypothetical, of human error, but only the pilot can put this information to use.

12
Conclusion

The related subjects of stalls and spins have been subjected to continuous attention from both pilots and aircraft designers throughout the history of manned flight. Since those early days of flight, the National Advisory Committee for Aeronautics and its successor, National Aeronautic and Space Administration, have conducted extensive research on the stall/spin characteristics of both military and general aviation airplanes. Unfortunately, the aerodynamic complexities associated with stalls and spins have defied the identification of a single solution to stall/spin accidents.

Although there has been a slow but steady reduction in stall/spin accidents, a significant decrease has not taken place. This in itself suggests that no single solution exists, but indicates instead that a combination of approaches must be considered. These approaches are stall awareness, pilot training, and spin-resistant airplanes.

Stall Awareness

In 1949, the Civil Aeronautic Board (CAB) eliminated the requirement that a pilot applicant demonstrate proficiency in spin entry and recovery during the certification flight test. Instead of spin training the CAB and the Federal Aviation Agency

113

shifted emphasis to prevention of and recovery from the stall condition. This stall-avoidance training was based on the fact that approximately 50 percent of stall/spin accidents occurred during take-off and landing. In these flight operations the altitude does not allow for spin recovery. It was determined that only 5 percent of the spin accidents occurred at altitudes sufficient to allow recovery. The stall-avoidance training as been effective as has been evidenced by the steady decline in such accidents over the years since the requirement was dropped.

Stall warning devices help prevent stall/spin accidents, but a warning still requires corrective action on the part of the pilot. To this end the FAA has clearly set forth the stall requirements (verbal, written, and flying) required for pilot certification. Pilots are required to know stall theory and to demonstrate entry and recovery from a variety of stall conditions; however, neither testing nor teaching take the element of surprise into consideration. Stall awareness, while essential to flight safety, has limited value.

Pilot Training

Although pilot training places great emphasis on stall recognition and recovery, spin entry and recovery is not required. This emphasis on stall awareness (as a prerequisite to spin avoidance) versus spin recovery was the subject of a congressional investigation in June 1980.

A hearing on spin-recovery training was given before the Subcommittee on Investigation and Oversight, Committee of Science and Technology, U.S. House of Representatives. Two key questions were addressed:

1. Should spin-recovery training be a requirement for obtaining a private pilot's license?
2. Would the requirement for spin-recovery training for a

private pilot's license significantly increase overall safety in general aviation?

Experts from all segments of the aviation community were divided on the issue. Therefore, no action was taken to make spin testing a part of private pilot certification.

The majority of stall/spin accidents take place at low altitudes (60 percent occur during take-off, landings, and go-arounds) where there is insufficient altitude for a developed spin prior to ground impact. Hence there is need to stress stall awareness in general, with particular attention to traffic pattern situations conducive to inadvertent stall entry. This cannot be done safely at pattern altitudes, but it is a subject for discussion during flight briefing and debriefing.

According to many experienced flight instructors, the most likely time for student pilots to inadvertently enter a spin is when practicing stalls. The student encounters a wind gust (surprise!) at stall entry, misuses the controls (usually the ailerons), and enters a spin. Unless the student has had in-depth spin-recovery instruction on the ground and in the air, the recovery is a hit-and-miss manipulation of the controls. Some students relate this event to their instructors, others never mention it, and a few simply quit flying. All of them experience an anxiety level never before known in their lives. Of course, there are those who do not recover, and this accounts for the high percentage of spin accidents in the "unknown" category (see Chapter 1, Table 1.3).

This is not meant to imply that spin-recovery instruction assures safe flight, but it is the quality of the instruction, both on the ground and in the air, that enhances flight safety. Again, be reminded that spin-recovery technique is of little value when at pattern altitude.

A flight instructor must remember that spin training requires an airplane designed for intentional spins. While most training aircraft are approved for intentional spins, most general aviation airplanes do not have such approval, and, as has been shown by NASA research, small changes in airplane configuration (CG, mass distribution, rigging) can significantly alter spin characteristics. Therefore, it is imperative that any airplane used for spin training have special maintenance attention (rigging, controls) and have detailed preflight inspection (nose gear, fuel, weight/balance, operation handbook).

Spin Resistance

Current FAA stall/spin certification standards address airplanes that spin; thus, the standards emphasize recovery characteristics. Provisions for a spin-proof airplane are included in the standards, but the nature of the regulation makes compliance a technically difficult process. It is easier for a designer to comply with the one-turn spin standard, but this produces a spinnable airplane.

Traditionally pilots know that most airplanes, when flown in a stall condition, show severe reductions in stability and control, a condition conducive to inadvertent spin entry. It can be logically concluded that spin prevention is dependent upon the inherent spin resistance of the airplane. The production of a spin-resistant airplane, however, depends on a means for FAA certification.

NASA, through flight tests of three light, general aviation airplanes, arrived at three criteria for development of standards needed to describe the desirable traits of a spin-resistant airplane. The basic idea was to maintain controllability in the stall regime so that the pilot retains full control of all three axes in terms of attitudes and rates. NASA suggested spin-resistance requirements be expressed strictly in terms of the airplane's controllability.

The first criterion requires maintenance of lateral control throughout stall. Because of loss of roll damping near stall contributes to wing drop, roll off, and autorotation, a spin-resistant airplane should permit wings-level flight (within 15 degrees) with no uncontrollable roll tendency when the pitch control is held full aft. The pilot should be able to change and control bank angle within a specific range (30 degree bank, right and left).

Because most stall/spin accidents result from inadvertent stall, a second criterion requires controllability even with uncoordinated control inputs. It is expected that a pilot responds to wing drop with opposite aileron (antispin), but spin resistance with the same aileron (prospin) would be a more definitive indicator of spin resistance for a worse case condition.

The third criterion addresses controllability after sustained control abuse represented by prolonged prospin input. A prospin input of 7 seconds duration was determined to be reasonable. However, a minimum time period of 4 seconds to reach 360 degrees heading change is needed to prevent disorientation. A rapid build-up of angular rotation produces pilot disorientation even though controllability can be maintained.

These criteria remain to be implemented, but the technology and knowledge exists. The modified leading edge configuration explored by NASA satisfy the criteria.

Summary

While there is a variation in degree, all flying involves a risk. It is the skilled pilot who is familiar with all the risks, and it is the wise pilot who acquires the knowledge and skills needed to minimize these risks. Such knowledge includes the aerodynamics of flight as well as the physiological limits of the pilot relative to a given flight operation.

The spin is not a normal flight operation for either the airplane or the pilot. Regardless of pilot skill and ability, the spin is a risk. To minimize the risk of a spin, the wise pilot acquires all the knowledge applicable to the spin. It is to this goal that this book is written.

CONCLUSION

BIBLIOGRAPHY

Beggs, Gene. "A Universal Spin Recovery Method." *Sport Aviation* 33 (August 1984):15–18.

Bowman, J. S., Jr. "Summary of Spin Technology as Related to Light General Aviation Airplanes." NASA TN D-6575, December, 1971.

Burk, S. M., Jr.; Bowman, J. S., Jr.; and White, W. L. "Spin-Tunnel Investigation of the Spinning Characteristics of Typical Single-Engine General Aviation Designs." NASA TP-1009, September, 1977.

Cessna Aircraft Company, *Spin Characteristics of Cessna Models 150, A150, 152, A152, 172, R172 & 177.* May, 1980.

Chambers, J. R. "Overview of Stall/Spin Technology." American Institute of Aeronautics and Astronautics. Paper 80–1580, August, 1980.

Christy, Joe. *The Guide to Single-Engine Cessnas.* Blue Ridge Summit, PA.: Tab Books Inc., 1979.

DiCarlo, D. J., and Johnson, J. L., Jr. "Exploratory Study of the Influence of Wing Leading-Edge Modifications on the Spin Characteristics of a Low-Wing Single-Engine General Aviation Airplane." AIAA Paper 79–1625, AIAA Aircraft Systems and Technology Meeting, New York, NY, August, 1979.

Federal Aviation Agency. Flight Training Handbook. AC 61–21, Washington, D.C.: GPO, 1965.

Hurt, H. H., Jr. *Aerodynamics for Naval Aviators.* Office of the Chief of Naval Operations, Aviation Training Division, NAVAIR 00–80T-80, January, 1965.

Hallion, R. P., *Designers and Test Pilots.* Alexandria, Va.: Time-Life Books, 1983.

Jones, G. M. "Vestibulo-ocular Disorganization in the Aerodynamic Spin." *Aerospace Medicine* 36(October 1965):976–83.

Kershner, W. K. *Basic Aerobatic Manual.* Ames, Ia.: Iowa State University Press, 1987.

Kirkham, W. R. "G Incapacitation in Aerobatic Pilots: A Flight Hazard." National Technical Information Service, FAA-AM-82-13, Springfield, Va., October, 1982.

Lan, C. E., and Roskam, Jan. *Airplane Aerodynamics and Performance.* Ottawa, Ks.: Roskam Aviation and Engineering Corporation, 1980.

Muller, Eric. "The Spin—Myth and Reality." *Sport Aerobatics* 10(August 1981):16–17.

Ralston, J. N. "Influence of Airplane Components on Rotational Aerodynamic Data for a Typical Single-Engine Airplane." AIAA Paper 83–2135, August, 1983.

Ranaudo, R. J. "Exploratory Investigation of the Incipient Spinning Characteristics of a Typical Light General Aviation Airplane." NASA TM X-73671, July, 1977.

Silver, B. W. "Statistical Analysis of General Aviation Stall/Spin Accidents." Society of Automotive Engineers Paper 760480, 1976.

"Staying Alive in an Airplane: A Game of Chance or Skill." *Aviation Safety* 3, no. 7 (July 1983):1–6.

Stewart, E. C.; Suit, W. T.; Moul, T. M.; and Brown, P. W. "Spin Tests of a Single-Engine High-Wing Light Airplane." NASA TP-1927, January, 1982.

Stinton, Darrol. *The Design of the Airplane.* New York: Van Nostrand Reinhold Co., 1983.

Stough III, H. P.; DiCarlo, D. J.; and Patton, J. M., Jr. "Flight Investigation of Stall, Spin, and Recovery Characteristics of Low-Wing, Single-Engine, T-Tail Light Airplane." NASA TP-2427, May, 1985.

Stough III, H. P.; Patton, J. M., Jr.; and Sliwa, S. M. "Flight Investigation of the Effect of Tail Configuration on Stall, Spin, and Recovery Characteristics of a Low-Wing General Aviation Research Airplane." NASA TP-2644, February, 1987.

Stough III, H. P.; Jordan, F. C., Jr.; DiCarlo, D. J.; and Glover, K. E. "Leading-Edge Design for Improved Spin Resistance of Wings Incorporating Conventional and Advanced Airfoils." SAE Paper 851816, October, 1985.

Stough III, H. P. and Patton, J. M., Jr. "The Effects of Configuration Changes on Spin and Recovery Characteristics of A Low-Wing General Aviation Research Airplane." AIAA Paper 79–1786, AIAA Aircraft Systems and Technology Meeting, New York, NY, August, 1979.

Van Patten, R. E. "Gs and Aerobatic Pilots." *Sport Aerobatics* 16(February 1987):6–9.

Whinney, J. E. "Gz-Induced Loss of Consciousness in Undergraduate Pilot Training." *Aviation Space and Environmental Medicine* 57(October 1986):997–999.

INDEX

TL 710 .D368 1989
DeLacerda, Fred.
Surviving spins